EXPLORING *the* OCEAN ENVIRONMENT

Michelle K. Hall
Science Education Solutions, Inc.

C. Scott Walker
Oregon State University

Jennifer A. Weeks
Science Education Solutions, Inc.

Larry P. Kendall
Science Education Solutions, Inc.

Jeff S. Jenness
Jenness Enterprises

THOMSON
BROOKS/COLE

Australia • Brazil • Canada • Mexico • Singapore
Spain • United Kingdom • United States

THOMSON
BROOKS/COLE

Exploring the Ocean Environment

Michelle K. Hall
C. Scott Walker
Jennifer A. Weeks
Larry P. Kendall
Jeff S. Jenness

Executive Editor: Peter Adams
Assistant Editor: Carol Benedict
Editorial Assistant: Anna Jarzab
Marketing Manager: Joe Rogove
Marketing Communications Manager: Bryan Vann
Content Project Manager: Belinda Krohmer
Creative Director: Rob Hugel
Art Director: Vernon Boes
Print Buyer: Nora Massuda
Permissions Editor: Roberta Broyer
Cover Designer: Denise Davidson
Cover Image: Corbis
Cover Printer: Courier-Kendallville
Printer: Courier-Kendallville

Printed in the United States of America
1 2 3 4 5 6 7 10 09 08 07 06

Library of Congress Control Number: 2006906786

ISBN 0495115061

Thomson Higher Education
10 Davis Drive
Belmont, CA 94002-3098
USA

Acknowledgments

Series Authors

Michelle K. Hall
Anne K. Huth
C. Scott Walker
Jennifer A. Weeks
Robert F. Butler
Jeff S. Jenness
Larry P. Kendall

Programmers

Jeff S. Jenness
Christian J. Schaller

Media Developers

Christian J. Schaller
Ted Stude

Technical Reviewers

Margaret Chernosky
Lydia Fox
Kathleen Friedman
Robert Kolvoord
Carla McAuliffe
Anne Ortiz
Alan Sills
Jessica Walker

Scientific Reviewers

Robert Butler
Peter Kresan
David Smith
Terry Wallace
Joseph Watkins

Technical Advisors

Tom Garrison
Cheryl Greengrove
Miles Logsdon
David Smith
Jim Washburne
Douglas Yarger

Proofreader

James Kevin

Data Providers

Robert Diaz
Virginia Institute of Marine Science
Don Pool
USGS Tucson Field Office
Pima County Department of Transportation Technical Services Division
Nancy Rabelais
Louisiana University Marine Consortium
Christopher Scotese
The Paleomap Project

Field Testers

Tekla Cook
Jo Dodds
Christine Donovan
Lillian Fox
Joshua Hall
Alan Kelley
Bob Kolvoord
Kathy Krucker
Kathy Likos
Daniel Montoya
Stephen Murray
Anthony Ochiuzzi
Graciela Rendon-Coke
Richard Spitzer
Anne Thames
Steven Uyeda
Margaret Wilch

Assistants

Tammy Baldwin
Christine Hallman
Sara McNamara
Marie Renwald
Megan Sayles

The authors would also like to thank the many scientists who took the time to learn about this project and share critical research data and expertise, and to the agencies and individuals that have given us permission to include their outstanding illustrations and photos.

SAGUARO
Science And GIS Unlocking Analysis & Research Opportunities
The SAGUARO Project

Michelle K. Hall, Director
Science Education Solutions, Inc.
Los Alamos Research Park
4200 West Jemez Drive
Synergy Center Suite 301
Los Alamos NM 87544-2587

saguaro@scieds.com

Science Education Solutions
understanding our world through science
http://www.scieds.com/saguaro

Table of contents

Introduction

Investigations

Getting started

Software needed

The investigations in this version of the GIS Investigations for the Earth Sciences series are intended for use on campuses with ArcGIS 9.x laboratory or site licenses. All necessary software and data files should be set up and ready for you to use on campus or lab computers. Your instructor will provide you with information on the location(s) of these computers, and where to access the data files for the investigations.

If you have a valid license for the student edition of ArcGIS 9.x, you may be able obtain the necessary data, additional software, and installation instructions to install and use these materials on your personal computer. Contact your instructor for information on where to obtain the required files.

In addition to ArcGIS 9.x, these materials require the following software. Most are available as free downloads from their respective publishers.

- Saguaro Tools for ArcGIS 9.x (contact your instructor)
- Web browser (Internet Explorer, Netscape, Opera, Mozilla, etc.)
- Windows Media Player
- Google Earth
- Adobe Reader

Attention course instructors!

The Instructor Guide (answer key), investigation data, additional software, installation instructions, troubleshooting guide, and other resources can be downloaded from the Instructor Resource Center at:

http://www.thomsonedu.com/earthscience/hall/ocean

You will need a user name and password, provided by the publisher, to access these files.

Mac OS X compatibility

At this time, ArcGIS 9.x is not compatible with any version of the Macintosh operating system. Users may get satisfactory, but slow, performance running ArcGIS under Virtual PC software on a Macintosh computer, but this is neither recommended nor supported.

What you need to know

The authors of this book assume that you have the following computer skills:

- Turning the computer on and, if necessary, logging in as a user.
- Navigating the file system to find folders, applications, and files.
- Launching applications and opening files.
- Opening, closing, moving, and resizing windows.
- Using tools, buttons, menus, and dialog boxes.

Other than these basic computer skills, students and instructors do not need prior experience using ArcGIS software. We do, however, strongly recommend that instructors run through each investigation before presenting it to students, to familiarize themselves with the data and techniques. We also strongly recommend that students and instructors:

- Be familiar with visual cues used in these materials.
- Read all instructions carefully.
- Pay attention to information provided in computer-screen shots and in the left-hand sidebar on each page.

What is the difference between ArcGIS, ArcView, and ArcMap?

For all practical purposes, you can think of these as three different names for the same geographic information system software.

Technically, ArcGIS is a family of related tools for managing geographic information systems on a variety of scales, ranging from Web servers to handheld devices.

One of these tools is a package for use on desktop computers, called ArcView. The ArcView 9.x software package consists of three components, each interacting with geographic data in a unique way:

- ArcCatalog — management
- ArcMap — viewing and analysis
- ArcToolbox — modifying data

In these materials, you will only be using the ArcMap component of ArcGIS 9.x.

Using ArcMap

Launching ArcMap and opening project files

- To launch the ArcMap application, click the Start button on the Windows Taskbar and choose **All Programs** > **ArcGIS** > **ArcMap** (ArcMap).
- If you see the ArcMap dialog box, choose **Browse for file** under the **An existing map** option.

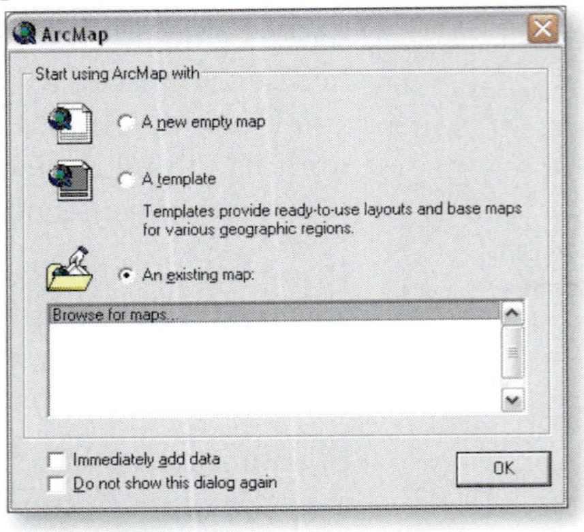

- Choose **File** > **Open**.
- Navigate to the appropriate unit data folder installed on your local hard drive or server and open it. (Your instructor or lab administrator can tell you where to find the investigation folders.)
- Locate the specified ArcMap document file and open it. (The .mxd file extension may or may not be visible, depending on how the computer has been set up.)

ArcMap shortcuts

In addition to the Start menu, an ArcMap shortcut may have been placed on the desktop or taskbar for your convenience.

Why doesn't the ArcMap startup dialog box appear?

If someone has checked the **Do not show this dialog again** checkbox, it will no longer appear each time you launch ArcMap.

Visual cues

Visual cues are used to make the investigation directions easier to follow.

- Text preceded by a computer symbol is an instruction—something to do on the computer.
- Names of tools or buttons are capitalized and are followed by a picture of that item as it appears on screen—for example, the Identify tool.
- The > symbol between boldface words or phrases in text indicates a menu choice. Thus, **File** > **Open…** means "pull down the File menu and choose Open…"

"File > Open…" means…

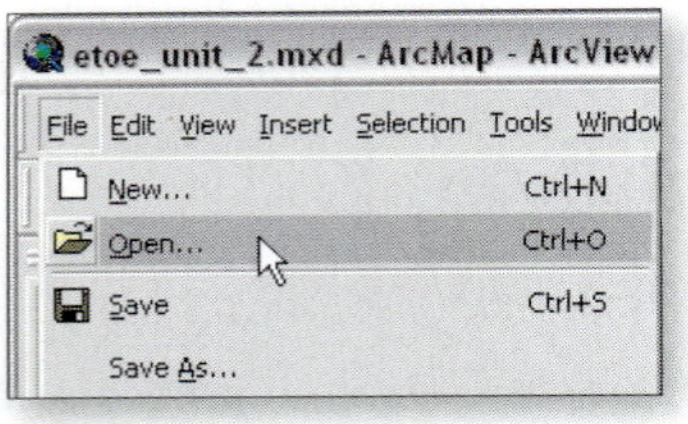

Sidebars contain important information!

The page sidebars contain useful information such as definitions, explanations, illustrations, examples, reminders, warnings, tips, and hints. If you are not sure what to do, look for help in the sidebar first.

Closing map files

When you have completed an investigation or must stop for some reason, choose **File** > **Exit** and click **No** when asked if you want to save your changes.

What if I click "Yes"?

Don't worry if you click **Yes** when you close a map file. The files have been locked to prevent you from accidentally modifying them.

The ArcMap User Interface

Title Bar
Shows the name of the current map file.

SAGUARO Tools
Tools provided by the SAGUARO Project for these materials. Note: This toolbar may appear in a different location.

Menu Bar
Provides menus for performing various operations.

Toolbars
Tools for manipulating and analyzing map data.

Table of Contents
A list of data frames and map layers, and controls for changing the map view.

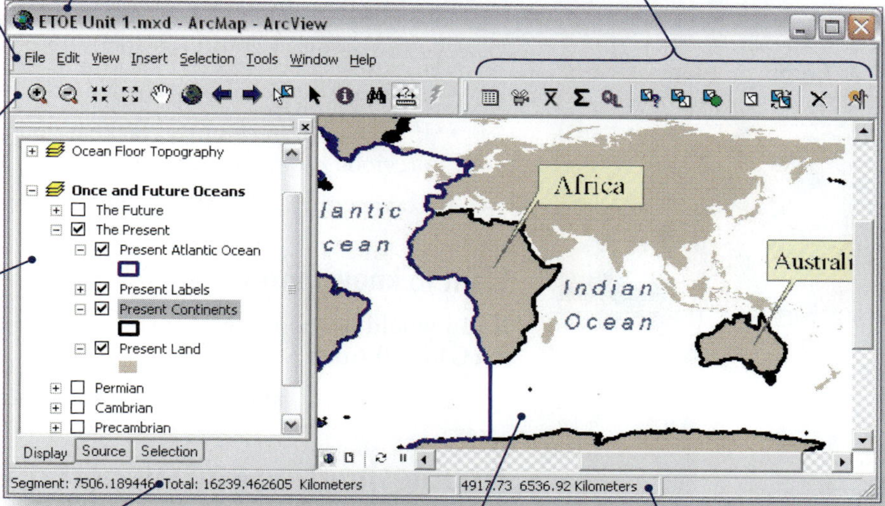

Status Bar
Displays tool descriptions, measurement results, and other information about the current operation.

Data Frame
The currently-selected map.

Coordinates
Displays the coordinates, in the currently-selected units, of the cursor position.

Basic display operations

Activating a data frame

To activate a data frame, right-click its name and choose Activate from the pop-up menu. The title of the activated data frame is highlighted bold.

Expanding and collapsing data frames

To expand a data frame and show its layers, click the expand box ⊞. To collapse a data frame and hide its layers, click the collapse box ⊟.

Selecting layers or layer groups

To select a layer or layer group, click the layer or layer group name. Selected layers or layer groups are highlighted. To select multiple layers or layer groups, hold down the control key while clicking additional names.

Expanding and collapsing layers or layer groups

To expand a layer or layer group, click the expand box ⊞. To collapse a layer or layer group, click the collapse box ⊟.

Turning layers or layer groups on and off

To turn a layer or layer group on, check ☑ the box in front of its name. To turn a layer or layer group off, uncheck the box ☐ in front of its name. If a layer is turned on but is not visible, it may be hidden behind another layer. Try turning off the layers *above* that layer in the Table of Contents.

Table of Contents

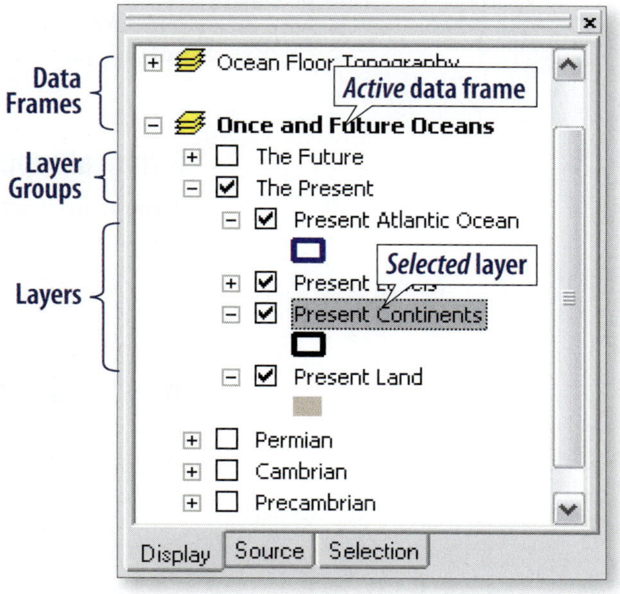

Zooming

ArcMap has tools for zooming—enlarging and reducing areas of the map—that work like the tools you have used in other applications.

- To zoom in on an area, click and drag diagonally with the Zoom In tool to outline the area on the map. When you release the button, the area you selected will rescale to fill the data frame window.
- To zoom out, click anywhere on the map with the Zoom Out tool.
- If you zoom in or out so far that you do not know where you are, undo previous zooms by clicking the Previous Extent button.
- To view the entire data frame, click the Full Extent button.

Zooming

The most efficient way to zoom in on a specific area is to drag a box around the area using the Zoom In tool. Drag diagonally from one corner of the area to the opposite corner, then release the button.

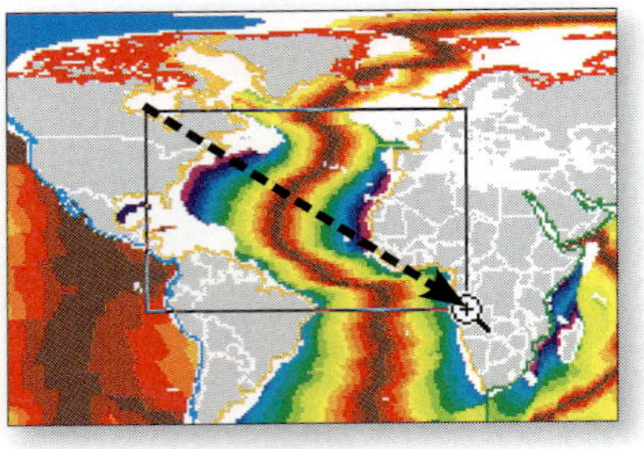

Want to know more?

If you would like to know more about using ArcGIS, download the **Guide to ArcGIS.pdf** from:

http://www.thomsonedu.com/earthscience/hall/ocean

Troubleshooting and support options

ArcGIS help

This module provides all of the directions you need to complete the investigations using ArcGIS 9.x. If you have other questions about the capabilities of ArcGIS, choose **Help > ArcGIS Desktop Help**.

Troubleshooting

The current version of the troubleshooting guide for this series (**Troubleshooting.pdf**) is available from the Instructor Resource Center at:

http://www.thomsonedu.com/earthscience/hall/ocean

Technical support

If you experience problems with these materials, contact the Thomson Learning technical support team at:

tl.support@thomson.com

1-800-4230563

Investigation data, updates, and resources

Investigation data, updates, and additional resources related to the *GIS Investigations for the Earth Sciences* series can be downloaded from:

http://www.thomsonedu.com/earthscience/hall/ocean

Working with large numbers

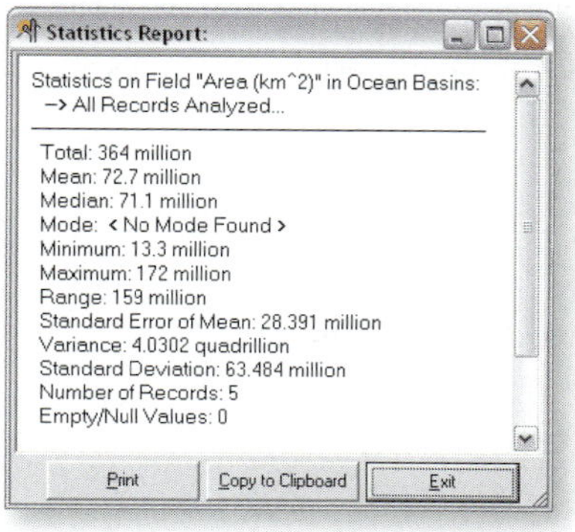

Some of the numbers you will work with in these investigations are quite large. When talking about the amount of water in the ocean or the energy of an earthquake or hurricane, you routinely use values in the billions or even trillions. Where possible, ArcGIS has been modified to make these very large and very small numbers easier to read. For example, in the Statistics Report window shown at left, the total area is given as 364 million, rather than 363958342077361 square kilometers.

Occasionally, you will need to convert millions to billions or thousands, or vice versa. For example, to convert the **Mean** value in the window at left from millions to billions, move the decimal point three places to the *left*. To go from millions to thousands, move the decimal three places to the *right*.

72700 thousand = **72.7 million** = 0.0727 billion

Rounding

Most of these numbers are approximations, so it does not make sense to be overly precise when you are calculating or recording them. Look at the number written below, and the place value of each of the digits. Face it—when you are talking about nearly 149 billion of something, who cares about hundred-thousandths, or even tens of millions?

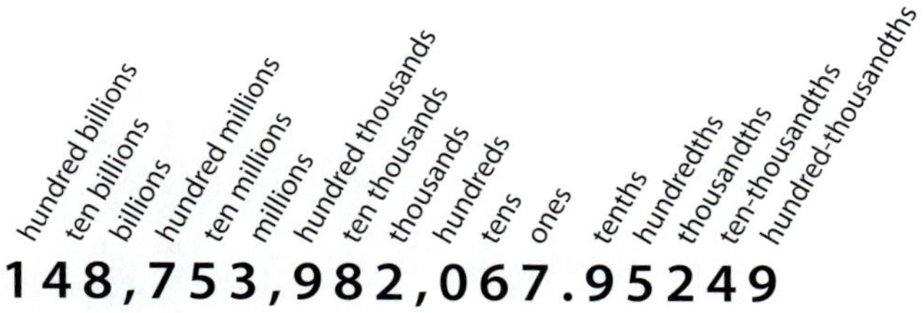

Throughout these investigations, you will be asked to round answers to a particular value and number of decimal places, such as "Round your answer to the nearest 0.1 million." Rounding numbers is simple, if you follow these steps. Examples are shown at the left.

- Look only at the numeral to the right of the place value you are rounding to. For example, when rounding to the nearest thousand, look only at the numeral in the hundreds place.
- If the numeral to the right is 0-4, do not change the number you are rounding to. If the number to the right is 5-9, add one to the number you are rounding to.
- Change whole numerals to the right of the place you are rounding to into zeros, and omit all unused decimal places.
- For any number less than 1, include a zero to the left of the decimal point. (Instead of *.79 billion*, write *0.79 billion.*)

Rounding examples

For example, if your number is

319,740,562.85

To round to the nearest ten million:

- Find the ten millions digit (1).
- Look at the number to its right (9). Because it is between 5 and 9, add one to the ten millions digit to make it 2.
- Change the whole numbers to the right of the ten millions digit to zeros and drop the decimal point and everything to its right. The result is **320,000,000**.

Rounding to the nearest...

...million (1,000,000) = **320,000,000** (adding 1 to 319 gives 320)

...hundred thousand (100,000) = **319,700,000**

...ten thousand (10,000) = **319,740,000**

...thousand (1,000) = **319,741,000**

...hundred (100) = **319,740,600**

...ten (10) = **319,740,560**

...one (1) = **319,740,563**

...tenth (0.1) = **319,740,562.9**

To round to the nearest 0.1 million:

- Find the 0.1 millions digit (7). This is also called the hundred thousands digit.
- Look at the number to its right (4). Because it is between 0 and 4, do not add one to the 0.1 millions digit.
- Insert the decimal point in the proper location. The result is **319.7 million**.

Rounding decimal fractions

Rounding decimals works the same way, except that you are rounding to tenths, hundredths, thousandths, and so on. Do not add zeros to the right of the decimal point. In other words, rounding 2.587 to the nearest tenth is 2.6, *not* 2.600.

Estimating percent area

You will occasionally be asked to estimate the percent area covered by land, ocean, or some other feature. This is a difficult skill for some people to master, but can be learned with practice.

Comparing to standards

One method of estimating coverage is to compare to visual standards. When estimating coverage you need to consider how the features are arranged.

Cloud cover exercise

Here is a simple activity that demonstrates the confusing nature of cover estimates.

- Take two full sheets of blue paper and one of white paper. The blue paper represents sky, and the white paper represents clouds.
- Cut the white sheet in half. Tear or cut the first half of the white sheet into large pieces and glue them onto one of the blue sheets without overlapping.
- Repeat the step above with the other half of the white sheet and the other blue sheet. This time, cut or tear the white sheet into small chunks before gluing them on.

In both cases, the cloud cover is 50 percent. Half of the blue sky is covered by white clouds, but the sheet covered by large clouds appears more open than the sheet covered by small clouds.

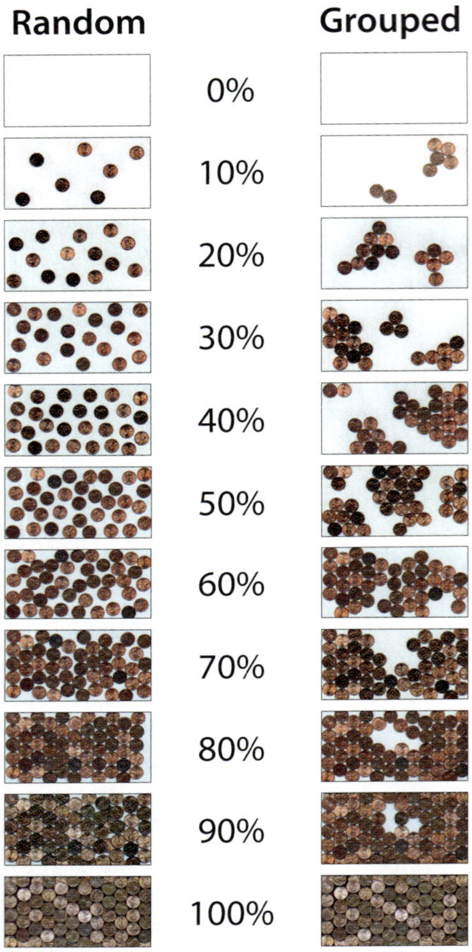

Gridding

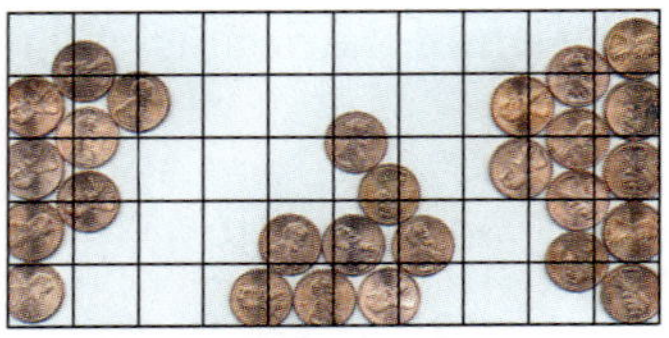

Another approach to estimating coverage is to divide the area up into a grid, either mentally or physically, and determine the number of grid squares that are at least half-covered. To find the percent coverage, calculate the ratio of covered squares to total squares and multiply by 100.

In the example at left, approximately 20 of the 50 squares are at least half covered.

$$20/50 \times 100 = 40\% \text{ coverage}$$

Philosophy & design

Philosophy

Thinking scientifically

An Earth scientist makes a living by observing and measuring nature. Whether recording and analyzing earthquakes or measuring subtle changes in sea surface temperature over many decades, a successful Earth scientist relies heavily on his or her ability to recognize patterns. Patterns in space and time are the keys to many of the great discoveries about how Earth works. The investigations in this series are designed to help you develop your ability to recognize and interpret nature's fundamental patterns by exploring recent scientific data using a computer and geographic information system (GIS) software.

What is a GIS?

GIS provides tools for organizing, manipulating, analyzing, and visualizing information about the world using digital maps and databases.

Most of these patterns are presented through maps, which are among scientists' most important tools. Maps allow you to visually explore spatial relationships between phenomena such as surface winds and ocean currents; natural features such as continents and ocean basins; and human features such as countries and cities. Behind each map layer is a table containing an extensive database of information about each feature in that layer. By carefully analyzing these data, you can identify patterns in the data that are difficult to discover through visual examination alone.

Planning to learn

Each unit of the series leads you through a well-tested learning process that builds upon your existing knowledge. Each unit begins with a warm-up exercise designed to stimulate your thinking about the major concepts presented in the unit and the key questions that motivate and guide scientific research. It will help you frame your own questions about the topic—questions that you may be able to answer for yourself as you learn more in later investigations.

In the first investigation, you will explore maps and data looking for patterns. As you examine these patterns, you should ask yourself questions such as:

- Where do they occur? (or not occur?)
- Why does this pattern occur here and not elsewhere?
- What might cause this pattern?
- What else is spatially associated with this phenomenon?
- Do these things usually occur together in the same places?
- How has this pattern changed spatially through time?

A brief reading provides key background information about scientific principles and concepts, and should help you begin to answer the questions raised earlier.

Finally, in one or more additional investigations, you apply your new knowledge to solve a particular problem. This helps you measure your understanding of the material and apply the concepts you have learned to a new location or situation.

GIS made easier

The purpose of these investigations is not simply to learn how to use GIS, but to use one as a tool to explore and learn about natural processes and features and how they relate to humans and human activities. For this reason, all of the data have been assembled into ready-to-use projects, and complex operations have been eliminated or simplified. Although it is helpful for you to have basic computer skills, you do not need experience with ArcGIS software to complete the investigations. The ArcMap user interface has been modified to simplify complex and repetitive processes. Directions for each task are provided in the text, so you will learn to use the tool as you explore with it. The investigations barely scratch the surface of the data that have been provided, and we encourage you to explore the data on your own.

Instructional Design: The 5-E Learning Cycle

This series was designed using the 5-E learning cycle model, which promotes inquiry and exploration as a process for learning science. The Learning Cycle, originally credited to Karplus and Thier (*The Science Teacher*, 1967) and later modified by Roger Bybee for the Biological Sciences Curriculum Study (BSCS) project, proposes that learning something new or understanding something familiar in greater depth involves making sense of both prior experience and firsthand knowledge gained from new explorations. The 5-E model divides learning experiences into five stages: Engage, Explore, Explain, Elaborate, and Evaluate. Each stage builds on the previous stages as you construct new understanding and develop new skills.

The 5-E Learning Cycle

Engage (Warm-Up)

This stage is designed to help you understand the learning task and make connections to past and present learning experiences. It should stimulate your interest and prompt you to ask your own questions about the topic.

Explore (Investigation)

In this stage, you investigate key concepts by exploring scientific, geographic, and economic data sets. You begin identifying patterns in the data and connecting them to Earth processes. This further stimulates curiosity and new questions develop. You may diverge from the written investigation to explore your own questions, continually building on your knowledge base. Through this process of questioning and exploration, you begin to formulate your understanding of basic concepts.

Explain (Reading)

This stage introduces you more formally to the lesson's scientific and geographic concepts. You should gain a better understanding of major concepts, acquire important terminology, and verify answers to questions or problems posed earlier. In addition, more abstract concepts not easily explored in earlier activities are introduced and explained.

Elaborate (Investigation)

In the Elaborate stage, you will expand on what you have learned and apply your newfound knowledge to different situations. You will test ideas more thoroughly and explore deeper relationships.

Evaluate (Wrap-up)

At the end of each unit, you will use your understanding of key concepts to propose explanations and solutions to local or regional problems.

Learning science with GIS

Geographic Information Systems (GIS) provide an ideal vehicle for learning topics in Earth and environmental sciences and helping you develop scientific problem-solving skills. Formerly limited to professionals with access to high-end computer workstations, today GIS is accessible to many, and is being used by students from elementary through graduate school. GIS has a number of advantages over traditional materials when used as an instructional tool. These include:

- **Data visualization**—GIS-based investigations allow you to identify and characterize relationships by manipulating multiple visual representations of data, including dynamic and customizable maps, tables, charts, and animations.
- **Data analysis**—Analytical tools enable you to quantify relationships within and among spatial data sets using database functions, statistical analyses, and spatial overlay operations.
- **Multimedia integration**—Other forms of digital information, including animations, video, audio, and digital stills, can be woven into GIS activities, greatly enriching and extending your learning experience.
- **Technology literacy and transferable skills**—The use of GIS promotes technology literacy and provides you with skills transferable to your own research, other course work, and the workplace.

GIS-based instructional materials have the potential to enhance your learning by reinforcing concepts through discovery and by improving problem solving, visualization, and computational skills.

Tear out this sheet for easy reference to common ArcMap operations.

GIS Investigations for the Earth Sciences

Quick Reference Sheet

The ArcMap User Interface

Title Bar
Shows the name of the current map file.

SAGUARO Tools
Tools provided by the SAGUARO Project for these materials. Note: This toolbar may appear in a different location.

Menu Bar
Provides menus for performing various operations.

Toolbars
Tools for manipulating and analyzing map data.

Table of Contents
A list of data frames and map layers, and controls for changing the map view.

Status Bar
Displays tool descriptions, measurement results, and other information about the current operation.

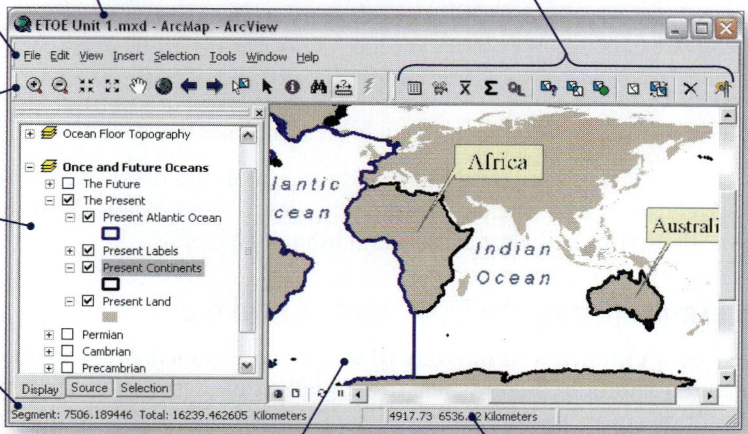

Data Frame
The currently-selected map.

Coordinates
Displays the coordinates, in the currently-selected units, of the cursor position.

Basic display operations

Activating a data frame

To activate a data frame, right-click its name and choose Activate from the pop-up menu. The title of the activated data frame is highlighted bold.

Expanding and collapsing data frames

To expand a data frame and show its layers, click the expand box ⊞. To collapse a data frame and hide its layers, click the collapse box ⊟.

Selecting layers or layer groups

To select a layer or layer group, click the layer or layer group name. Selected layers or layer groups are highlighted. To select multiple layers or layer groups, hold down the control key while clicking additional names.

Expanding and collapsing layers or layer groups

To expand a layer or layer group, click the expand box ⊞. To collapse a layer or layer group, click the collapse box ⊟.

Turning layers or layer groups on and off

To turn a layer or layer group on, check ☑ the box in front of its name. To turn a layer or layer group off, uncheck the box ☐ in front of its name. If a layer is turned on but is not visible, it may be hidden behind another layer. Try turning off the layers *above* that layer in the Table of Contents.

Table of Contents

Data Frames

Layer Groups

Layers

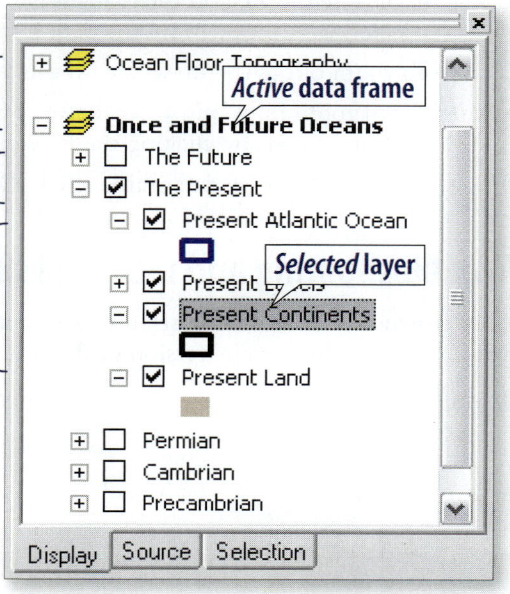

Tear out this sheet for easy reference to common ArcMap operations.

Opening ArcMap document (.mxd) files

At the beginning of each investigation, you will launch the ArcMap application and use it to open an ArcMap document file.

The instruction will look like this:

Launch ArcMap, and locate and open the **etde_unit_1.mxd** file.

etde_unit_1

ArcMap document files end with an .mxd file extension. Depending on how your computer has been set up, file extensions may not be visible, so the above file may appear as simply **etde_unit_1**. (Note: We will also refer to ArcMap document files simply as *map files*.)

Launching ArcMap

To launch the ArcMap application, click the Windows Start menu and choose **All Programs** > **ArcGIS** > **ArcMap**.

Shortcut to ArcMap.exe

Note: If there is an ArcMap shortcut icon on the desktop, Start menu, or taskbar, you can also use it to launch ArcMap. It may or may not show the .exe file extension. Your instructor or lab administrator may provide you with alternate directions for launching ArcMap.

Locating and opening the EXXX_Unit_X.mxd file

Choose **File** > **Open...** and navigate to where the specified ArcMap document file is located. Your instructor or lab administrator should tell you where the investigation data files are located. They will be in folders named with the module abbreviation and unit number, such as ETDE_Unit_1. Remember, you may not see the .mxd file extension.

Important note: Do not call or e-mail technical support if you cannot find an ArcMap document file. Your instructor or lab administrator can tell you where to find it!

Zooming

ArcMap has tools for zooming—enlarging and reducing areas of the map—that work like the tools you have used in other applications.

- To zoom in on an area, click and drag diagonally with the Zoom In tool to outline the area on the map. When you release the button, the area you selected will rescale to fill the data frame window.
- To zoom out, click anywhere on the map with the Zoom Out tool.
- If you zoom in or out so far that you do not know where you are, undo previous zooms by clicking the Previous Extent button.
- To view the entire data frame, click the Full Extent button.

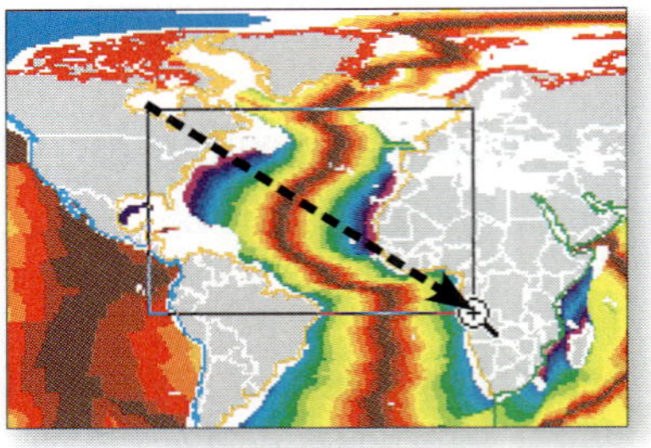

Troubleshooting and technical support

Your instructor or lab administrator may be able to help with common problems. If necessary, they can provide you with a copy of the current version of the troubleshooting guide for this series, **Troubleshooting.pdf**.

If you continue experiencing problems using these materials, contact the Thomson Learning technical support team at:

tl.support@thomson.com *or* **1-800-423-0563**

Unit 1
The Ocean Basins

In this unit, you will

- *Track changes in the world's continents and ocean basins from 750 million years in the past to 250 million years in the future.*
- *Contrast the extent and age distribution of ocean floor and continental crust and hypothesize about their origins.*
- *Investigate the origin of deep ocean basins.*
- *Create bathymetric profiles of an ocean basin at various resolutions.*
- *Examine surface features that reveal the dynamic nature of the ocean basins.*

NASA

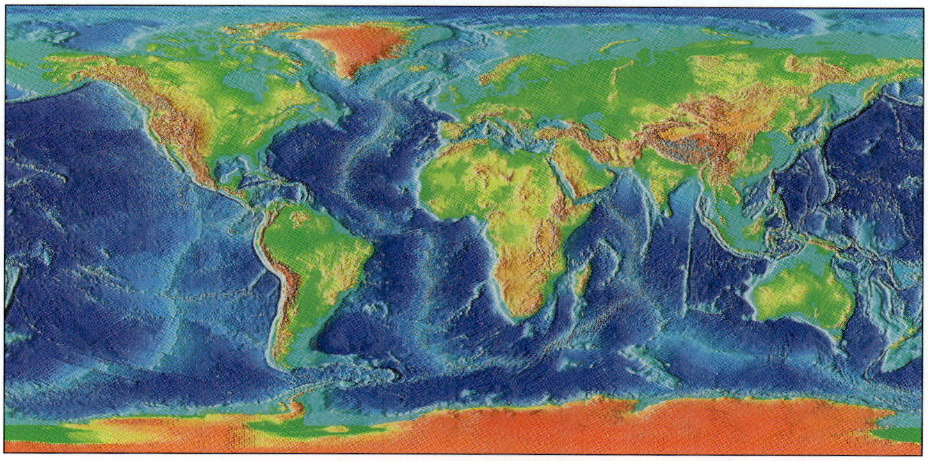

Surface features of ocean basins show that the ocean floor is constantly changing.

Warm-up 1.1

Oceans on other planets?

In early 2004, twin NASA rovers *Spirit* and *Opportunity* landed on Mars and found evidence that large quantities of water existed on its surface in the distant past. How would an ocean on Mars have been different from one on Earth, and what caused it to disappear?

The once and future ocean

Earth is unique in our solar system for its vast oceans of water. With over 70 percent of its surface covered by water, Earth is often called the *water planet.* Without this water, Earth would not have the diversity of life forms, the continents we live on (it takes water to create most continental rocks), or the atmosphere that protects us. Have you ever wondered where this water came from or how and when the ocean basins formed? In this unit, you will explore these questions. To start, think about these questions and write your best answers based on your current understanding of Earth. Don't be afraid to guess based on what you already know.

1. Consider that water is made up of oxygen and hydrogen atoms. Where do you think Earth's vast supply of water may have come from? Explain.

If Earth's surface were entirely smooth, with no mountains or basins, water would cover the entire planet.

2. What structural changes that occur in and on the surface of the Earth over time might cause Earth to have deep basins surrounded by high continents to hold the ocean water?

The everlasting ocean

And thou vast ocean, on whose awful face
Time's iron feet can print no ruin-trace.

Robert Montgomery, *The Omnipresence of the Deity*

Poets and writers are fond of saying that the oceans are everlasting, that a stroll along a sandy beach provides a glimpse into Earth's past and a look into its future. True, the oceans have been around for a long time—at least 3.8 billion years.

They will probably be around until the Sun expands and boils them away billions of years in the future. However, have they changed at all over time? For now, consider the poet's statement that the oceans are everlasting and unchanging. How true is this? You will use ArcMap to explore this question.

The Sun's lifetime

Astronomers estimate that the Sun's life span is about 10 billion years, and that it is currently around 4.6 billion years old.

Moving continents

- Launch ArcMap, and locate and open the **etoe_unit_1.mxd** file.

Refer to the tear-out Quick Reference Sheet located in the Introduction to this module for GIS definitions and instructions on how to perform tasks.

- In the Table of Contents, right-click the **Once and Future Oceans** data frame and choose Activate.
- Expand the **Once and Future Oceans** data frame.

This data frame shows the present locations of today's continents and oceans. However, the continents and oceans have not always been in these locations, nor will they be in the future. In this investigation, you will trace the movements of three continents from 750 million years ago (750 Ma) to 250 million years in the future (+250 m.y.). (See sidebar for an explanation of the abbreviations.)

- Using the Select Elements tool, click on the black North America label (NA) in the row of Present labels (Figure 1 at left). All of the continent labels, such as NA, are color coded according to their ages.
- Drag the selected label and place it in the middle of the continent it represents (North America) on the ArcMap screen.
- Repeat this process for the black Africa and Australia labels.
- Turn off and collapse the **Present** layer group.
- Turn on and expand the **Permian** layer group. The **Permian Labels**, **Permian Continents (250 Ma**, **Permian Land (250 Ma)**, and **Permian Oceans (250 Ma)** layers should be turned on.

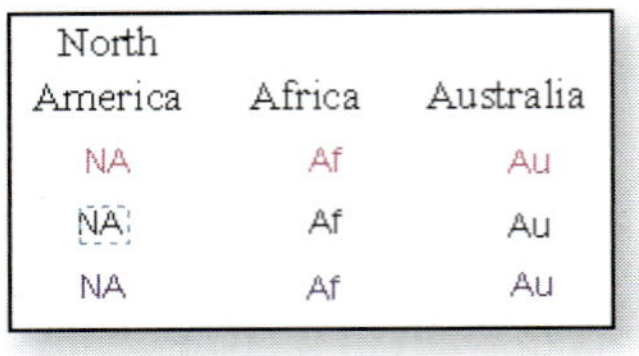

Figure 1. A blue, dashed line appears around the outside of a label when it is selected.

What are Ma and m.y.?

The Greek prefix *Mega-* represents *one million* in the metric system. The Latin word for year is *annum*. As used in Earth science, Ma stands for *Mega-annum*, or *millions of years before present*.

The abbreviation m.y. stands for *millions of years*. A + sign before the number represents millions of years *in the future*. (For example, +250 m.y. means *250 million years in the future*.)

During the Permian period, around 250 Ma (Figure 2 at left), the world looked very different than it does today. The **Permian Continents (250 Ma)** layer shows the approximate boundaries of the continents during the Permian period.

- Select and drag the blue North America (NA) label to the approximate center of that continent during the Permian period (250 Ma).
- Repeat this process for the blue Africa and Australia labels.
- Turn off and collapse the **Permian** layer group.
- Repeat this process to plot the location of the continents for the **Cambrian (500 Ma)**, **Precambrian (750 Ma)**, and **Future (+250 m.y.)** layer groups.
- Turn off and collapse all layers.
- Turn on and expand the **Present** layer group.
- Turn off all layers within the **Present** layer group except **Present Land**.

The purpose of the next exercise is to track the movement of North America through time.

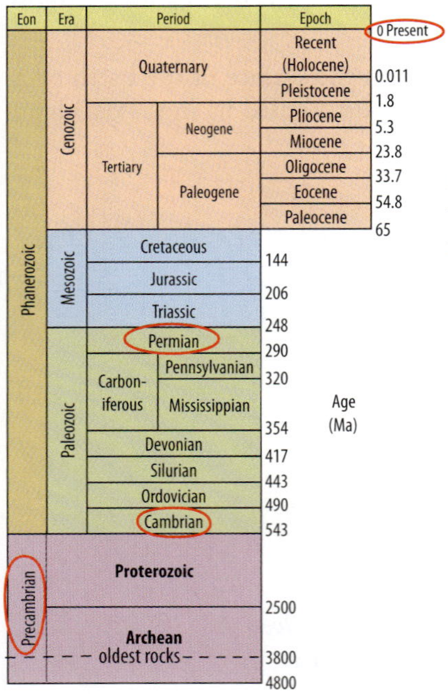

Figure 2. The geologic time scale. The time divisions are based on major changes in Earth's flora and fauna over time.

3. On Map 1, mark and label the location of North America during each time period. Draw a line with arrows showing the direction of movement of North America, from Precambrian to Cambrian to Permian to Present to Future.

Map 1 — Moving continents (750 Ma to +250 m.y.)

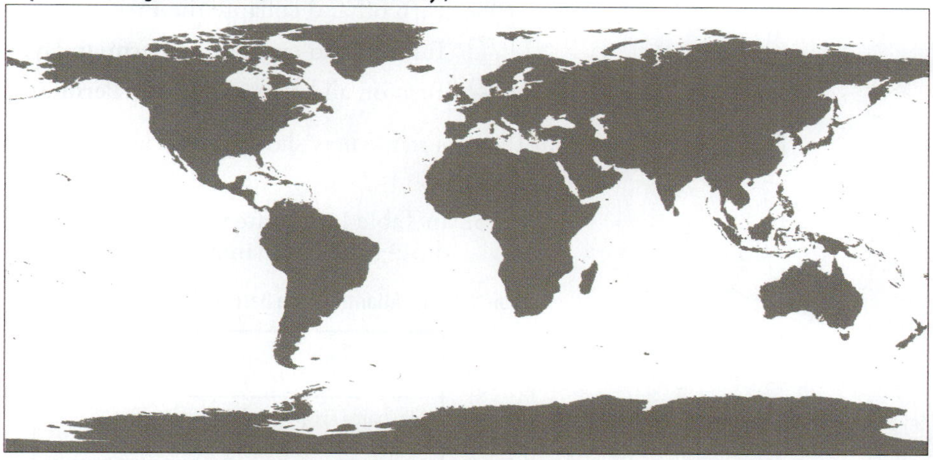

Repeat the process, labeling the locations of Africa and Australia during each time period on the map, and plotting the movement of each continent through time. (Hint: When you plot the path of Australia, remember that the left edge of the map connects with the right edge of the map because Earth is a sphere. So, your path may have to break when it reaches the edge of the map.)

speed — the ratio of distance to time. Mathematically,

$$\text{speed} = \frac{\text{distance}}{\text{time}}$$

4. Because the time intervals between labels are equal — 250 million years — the distance between labels is directly related to a continent's speed. To understand this concept, think of two cars that drive for an hour. The car that has travelled the longest distance in an hour had a higher speed than the other car. Choose and circle the answer that best completes the following statements.

   a. The average speed of the North American continent was faster / slower between the Precambrian and Cambrian periods than between the Cambrian and Permian periods.

   b. The average speed of the North American continent was faster /slower between the Cambrian and Permian periods than between the Permian and Present periods.

   c. The speed of the North American continent has been constant/ changing through time.

You are finished with the labels. To delete them, choose **Edit ▸ Select All Elements** followed by **Edit ▸ Delete.** (If parts of the labels still appear, they will disappear in the next step.)

To view animations of the continents moving over time, see the sidebar.

Moving continent animations

To view animations of the continents moving over time, click the Media Viewer button and choose either of the following movies from the media list.

- **Atlantic Basin History**
  Shows the changes in the Atlantic Ocean basin.
- **Moving Continents**
  Shows what Earth may have looked like as the continents moved.

USGS

The changing Atlantic

Next, you will focus on changes in the Atlantic Ocean over time.

Turn on the **Present Land** and **Present Atlantic Ocean** layers.

Turn off all other layers within the **Present** layer group.

This map shows the Atlantic Ocean as it exists today, situated between North and South America to the west and Europe and Africa to the east. We will use its location between these continents as a definition of the Atlantic Ocean.

Turn off and collapse the **Present** layer group.

Turn on and expand the **Permian** layer group.

Turn on all layers within the **Permian** layer group.

The data frame now shows Earth's land and ocean areas during the Permian period, 250 Ma.

5. In Table 1, indicate whether an earlier Atlantic Ocean did or did not exist during the Permian period.

Table 1 — The Atlantic Ocean basin (250 Ma to +250 m.y.)

Time	Permian *250 Ma*	Present	Future *+250 m.y.*
Probable existence of Atlantic Ocean		exists	

Turn off and collapse the **Permian** layer group.

Turn on and expand the **The Future** layer group.

Turn on all layers within the **The Future** layer group.

This map shows one possible arrangement of the continents +250 m.y. Although the past locations and extents of continents are fairly well known, the future locations and extents of the continents and oceans are not known. The locations shown here represent just one of several possible scenarios for the future Atlantic Ocean basin.

6. In Table 1, indicate whether this model suggests that the Atlantic Ocean will exist or will not exist +250 m.y.

7. Based on Table 1, describe in words what happens to the Atlantic Ocean basin from 250 Ma to +250 m.y.

This investigation began with a quote from the poet Robert Montgomery that suggested that the oceans are everlasting and unchanging.

8. Why do you think the oceans *seem* everlasting and unchanging?

9. Based on what you have learned, what argument would you use to support the statement that ocean basins are not everlasting and unchanging.

Throughout this unit, you will explore the evidence for and the driving forces behind the motion of the continents and the opening and closing of ocean basins. In later units, you will be challenged to think about the effects these changes have on ocean and atmospheric circulation, climate, marine productivity, and biological evolution and diversity.

Quit ArcMap and do not save changes.

Investigation 1.2

Changing oceans

According to the fossil and rock records, oceans have existed on Earth for at least 3.8 billion years. It is easy to think that they are permanent features, ancient and unchanging. However, the only thing that is truly constant about ocean basins—and continents—is that they are always moving and changing. New oceans are born, and existing oceans change size and shape or disappear entirely. How does this happen, and over what time scale? In this activity, you will explore evidence to answer these questions.

Using ocean-floor age data, you will identify major transformations that have occurred in Earth's ocean basins and when they took place. Differences in the ages of the ocean basins and continents provide clues to understanding when and how they formed.

Ocean-floor age

By measuring the ages of ocean-floor rocks both directly and indirectly, scientists have produced age maps of most of the ocean basins.

Launch ArcMap, and locate and open the **etoe_unit_1.mxd** file.

Refer to the tear-out Quick Reference Sheet located in the Introduction to this module for GIS definitions and instructions on how to perform tasks.

In the Table of Contents, right-click the **Ocean-Floor Age** data frame and choose Activate.

Expand the **Ocean-Floor Age** data frame.

What are Ma and m.y.?

The Greek prefix *Mega-* represents *one million* in the metric system. The Latin word for year is *annum*. As used in Earth science, Ma stands for *Mega-annum*, or *millions of years before present*.

The abbreviation m.y. stands for *millions of years*. A + sign before the number represents millions of years *in the future*. (For example, +250 m.y. means *250 million years in the future*.)

The **Detailed Ocean-Floor Age** layer shows the ages of the rocks that form the present ocean floor. Each colored band represents a time span of 10 million years.

1. Use the legend for the **Detailed Ocean-Floor Age** layer to complete Table 1.

Table 1 — Ocean-floor age

Rocks	Color	Age *Ma*
Youngest		
Oldest		

2. Describe the pattern of rock ages on the ocean floor. Do the ages of the rocks increase or decrease in a pattern, or are the ages disorganized and jumbled?

Turn on the **Detailed Continent Age** layer.

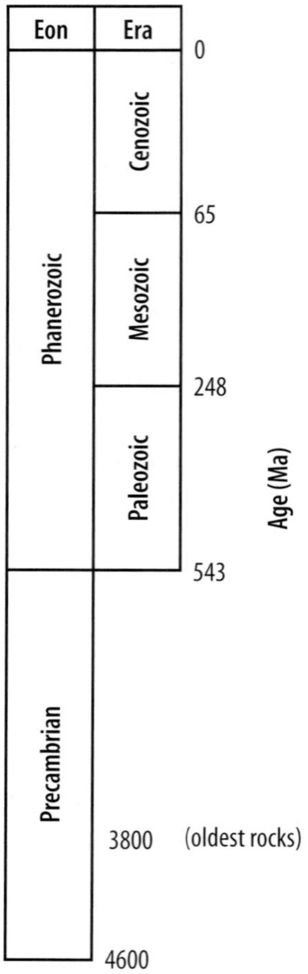

Figure 1. Geologic time scale (drawing not to scale).

This layer displays the ages of surface rocks on the continents, in periods of time that correspond to the major eras of geologic time (Figure 1 at left). A geologic era is a period of time that stands out from surrounding periods of time because of a significant change or event. Each of the different eras shown in Figure 1 are periods of time between mass extinctions.

3. Complete Table 2 using the legend for the **Detailed Continent Age** layer.

Table 2 — Continental rock age

Rocks	Color	Era / Eon	Age *Ma*
Youngest			
Oldest			

4. Do the ages of rocks on the continents follow a pattern similar to the one you saw in the ocean-floor rocks? Explain.

5. How many times older are the oldest continental rocks than the oldest rocks that form the ocean floor? (Hint: Divide the age of the oldest continental rocks in millions of years by the age of the oldest ocean-floor rocks in millions of years.)

Comparing oceanic and continental rocks

Next, you will compare the percentages of the ocean floor and the continents that are covered by young and old rock. You will use simplified versions of the ocean-floor and continental age layers that sort the rock as either young, created in the Cenozoic era (0–65 Ma); or old, created prior to the Cenozoic era (64–3800 Ma).

- Turn off and collapse the **Detailed Ocean-Floor Age** layer.
- Turn off and collapse the **Detailed Continent Age** layer.
- Turn on the **Ocean-Floor Age** layer.
- Select the **Ocean-Floor Age** layer.

Now you will determine the percentage of young rocks in the ocean floor.

- Click the Select By Attributes button.
- To select young ocean-floor rocks, query the **Ocean-Floor Age** layer for **("Age Category" = 'young')** as shown in steps 1–6. The query will actually read:

  ("AGE_CLASS" = 'Young').

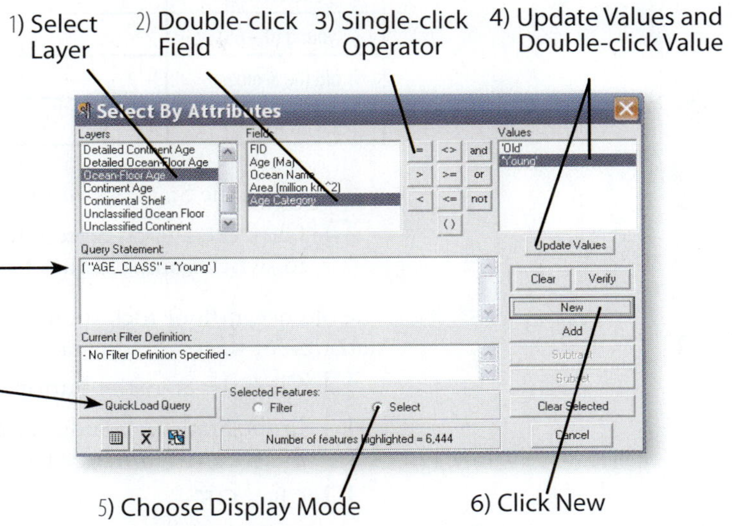

Read query statement here as you enter it.

QuickLoad Query

- Click the QuickLoad Query button and select the **Young Ocean Floor** query.
- Click **OK**.
- Click **New**.

- If you have difficulty entering the query statement correctly, refer to the **QuickLoad Query** described at left.

The young rocks of the ocean floor should now be highlighted. Next you will calculate statistics for the selected data.

- Click the Statistics button in the Select By Attributes window.
- In the Statistics window, calculate statistics for **only selected features** of the **Ocean-Floor Age** layer, using the **Area (million km^2)** field.

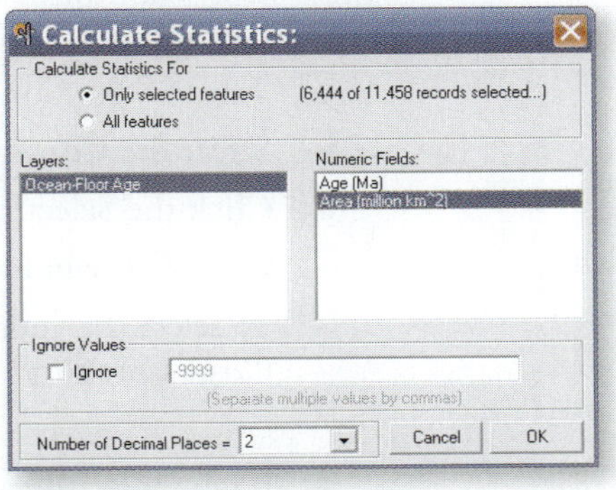

- Click **OK**.

The total area of the ocean floor that is covered by young rock is reported in the Statistics window as the **Total**.

6. Round the total area of young ocean-floor rock to the nearest whole number and record it in Table 3 on the following page.

Table Hint

This table will not be entirely filled in until you complete the instructions in Question 11.

Table 3 — Area and percent area of ocean-floor and continental rocks

Age	Ocean Floor		Continents	
	Area *million km²*	% Area	Area *million km²*	% Area
unclassified	32		2	
young (0 – 65)				
old (66 & older)				
Total		100.0		100.0

- Close the Statistics window.
- Click the Switch Selection button in the Select By Attributes window to switch your selection from the young to the old ocean-floor rock.

The old ocean-floor rock should now be highlighted. Next you will calculate statistics for the selected data.

- Click the Statistics button in the Select By Attributes window.
- In the Statistics window, calculate statistics for **only selected features** of the **Ocean-Floor Age** layer, using the **Area (million km^2)** field.
- Click **OK**.

7. Round the total area of old rock in the ocean floor to the nearest whole number and record it in Table 3.

- Close the Statistics window.
- Click **Clear Selected** in the Select By Attributes window.
- Close the Select By Attributes window.
- Turn off the **Ocean-Floor Age** layer.

These steps will now be repeated to obtain the area of the *continents* covered by young and old rock.

- Turn on the **Continent Age** layer.
- Select the **Continent Age** layer.
- Click the Select By Attributes button.
- Click **Clear** in the Select By Attributes window to erase the previous query.
- To select the young continent rocks, query the **Continent Age** layer for **("Age Category" = 'Young')**, and click **New.** The query will actually read:

  ("AGE_CATE_1" = 'Young').

- If you have difficulty entering the query statement correctly, refer to the **QuickLoad Query** described at left.
- Click the Statistics button in the Select By Attributes window.
- In the Statistics window, calculate statistics for **only selected features** of the **Continent Age** layer using the **Area (million km^2)** field.
- Click **OK**.

8. Round the total area of young continental rock to the nearest whole number and record it in Table 3.

QuickLoad Query

- Click the QuickLoad Query button and select the **Young Continent** query.
- Click **OK**.
- Click **New**.

- Close the Statistics window.
- Click the Switch Selection button in the Select By Attributes window to switch your selection from the young to the old continental rock.

The old continental rock should now be highlighted. Next you will calculate statistics for the selected data.

- Click the Statistics button in the Select By Attributes window.
- In the Statistics window, calculate statistics for **only selected features** of the **Continent Age** layer using the **Area (million km^2)** field.
- Click **OK**.

9. Round the total area of old continental rock to the nearest whole number and record it in Table 3.

- Click **Clear** in the Select By Attributes window to erase the previous query.
- Close the Statistics and Select By Attributes windows.

You can use this information to determine the percentage of ocean floor and continents represented by each age category. The following steps will help you complete Table 3.

Calculating percentages

Example: To calculate the percent of unclassified ocean-floor rock, divide the unclassified area by the total area and multiply by 100. Note: The area units — (million km^2 in this case) — must be the same.

Age *Ma*	Ocean Area *millions km^2*
unclassified	32
young (0-65)	
old (66 & older)	
Total	306

% Unclassified Ocean Floor

$$= \frac{32 \text{ million km}^2}{306 \text{ million km}^2} \times 100 = 10.5\%$$

10. Add the areas of unclassified, young, and old ocean-floor rock and record the total area in Table 3. Repeat this process to record the total area of continental rock.

11. Using the method shown at left, calculate the percentages of unclassified, young, and old ocean-floor rock, and record your results in Table 3. Repeat this process for the continental rock. (Note: Your computer probably has a calculator tool to help you with these calculations.)

12. According to Table 3, which contains a larger percentage of younger rock (less than 65 Ma) — the ocean floor or the continents?

Figure 2 (on the following page) shows the amount, in square kilometers, of rocks of different ages in today's ocean floor. For example, about 17 million km^2 of ocean floor is between 0 and 5 Ma.

13. According to Figure 2 on the following page, what is the area (million km^2) of the ocean floor that is between 75 and 80 Ma?

14. Describe any trend in the area data from young to old ocean floor (from left to right in Figure 2 on the following page).

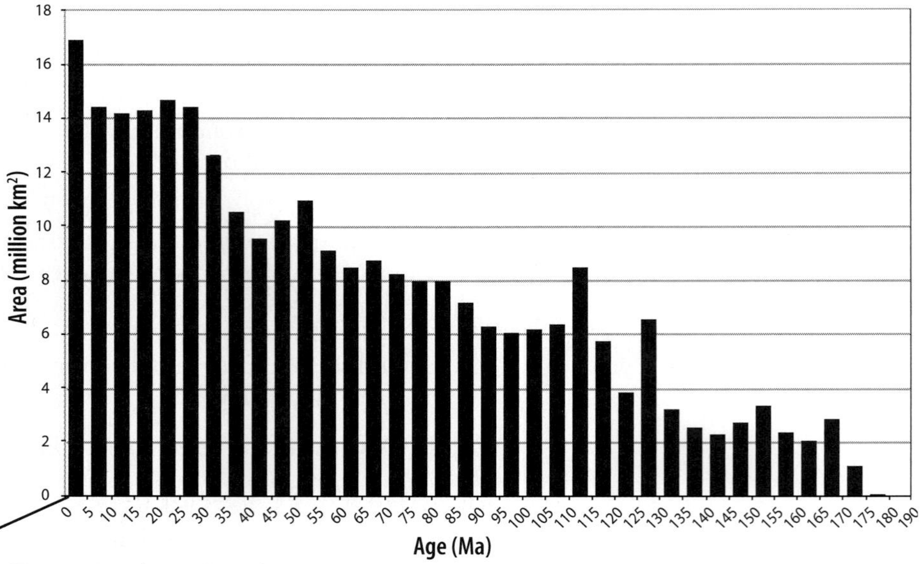

The present age is 0 million years ago

Figure 2. Area of ocean-floor rock versus age.

Scientists have contemplated the reason for the difference in the amount of older and younger ocean-floor rocks. We know that Earth has not grown in size in the past 4.6 billion years. So, if new crust is added in one place by some process, it must be destroyed somewhere else. Earth's systems and processes are connected; if the rate or amount of one process increases or decreases, it causes changes in other processes. We also know that the ocean basins have existed for 3.8 billion years, but the oldest rocks of the ocean floor are a mere 190 Ma. (Older ocean crust has been preserved on some continents.) What explains these observations? You will explore this next.

Formation of the Atlantic and Pacific Oceans

The ages of ocean-floor rocks can be used to determine when individual ocean basins formed and the rate at which they have expanded. Next, you will investigate the ages and expansion rates of the Atlantic and Pacific Ocean basins.

- Turn off the **Continent Age** layer.
- Click the QuickLoad button.
- Select **Spatial Bookmarks,** choose **Atlantic Basin**, and click **OK**.
- Turn on and expand the **Detailed Ocean-Floor Age** layer.

15. Remember that the Atlantic Ocean's southeastern limit is the southern tip of Africa. Where is the youngest and oldest ocean floor in the Atlantic?

16. Based on the oldest ocean floor in the Atlantic, when did the Atlantic

Ocean basin first begin to open up?

17. Did the North Atlantic Ocean basin open at the same time as the South Atlantic Ocean basin? Explain. (Hint: Compare the ages of the oldest rocks in each basin.)

- Turn on the **Ridges** layer.
- Click the QuickLoad button.
- Select **Spatial Bookmarks,** choose **North Atlantic**, and click **OK**.
- Locate the ocean floor that was created in the last 10 million years.
- Use the Measure tool to determine the total width of new ocean floor that has been created in the North Atlantic Ocean basin over the last 10 million years. Use of the Measuring tool is described in detail in Figure 3, at left.

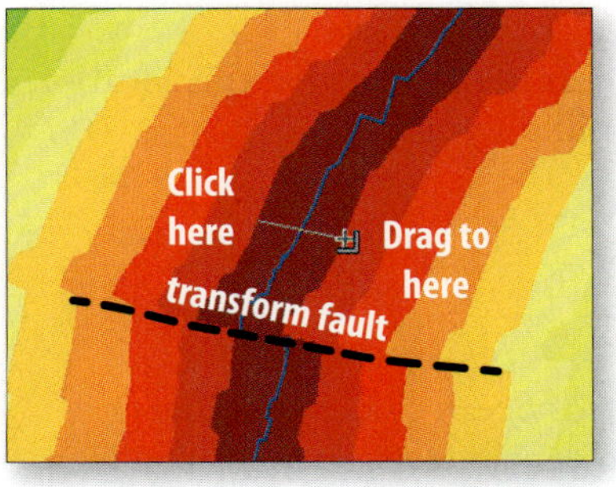

Figure 3. Measuring total width of new ocean floor created. To measure the width of the 0–10 million year band (dark red):

- Click on one side of the band with the Measuring tool.
- Drag across to the other side of the band, parallel to the transform faults along the ridge.
- Double-click to stop measuring.

The total distance measured is shown at the bottom of the screen.

18. What is the total width, in kilometers, of new ocean floor that has been created in the North Atlantic Ocean basin in the last 10 million years?

19. Calculate the rate of ocean-floor spreading in the North Atlantic Ocean basin by dividing the width of new ocean floor that you just measured by the time required to create it (10 million years). Convert your result to centimeters per year. (See *Converting units* in sidebar.)

Converting units

To convert from width to rate, divide the width in kilometers by 10 million years:

$$\frac{\text{width (km)}}{\text{10,000,000 yr}} = \text{rate (km/yr)}$$

To convert the rate from kilometers per year to centimeters per year, multiply by 100,000:

$$\frac{\boldsymbol{X}\ (\cancel{\text{km}})}{\text{1 yr}} \times \frac{\text{100,000 cm}}{1\ \cancel{\text{km}}} = \text{rate (cm/yr)}$$

- Click the QuickLoad button.
- Select **Spatial Bookmarks,** choose **South Atlantic**, and click **OK**.
- Use the Measure tool to determine the total width of new ocean floor created in the South Atlantic Ocean basin over the last 10 million years.

20. What is the width, in kilometers, of new ocean floor that has been created in the South Atlantic Ocean basin in the last 10 million years?

21. Which part of the Atlantic Ocean basin is spreading at a faster rate—the North or South Atlantic?

22. Using the measurement from Question 20, calculate the spreading rate of the South Atlantic Ocean basin, in centimeters per year.

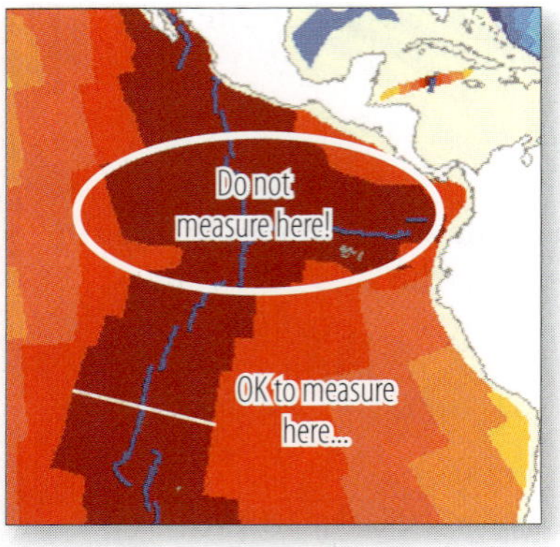

Figure 4. Caution: Do not measure the width of new ocean floor in the region of the eastern Pacific Ocean shown above. Two ridges (blue lines) intersect in this area, making it difficult to measure either one separately.

Click the Full Extent button to view the entire map.

Now look at the Pacific Ocean. A disadvantage of flat maps of Earth is that it is not obvious that a feature on the left edge of the map continues on the right edge. Here, the Pacific Ocean appears split into two separate parts—off the coast of North and South America on one side and off the coast of Asia and Australia on the other side. Be sure to examine the entire Pacific Ocean.

23. How old are the oldest rocks of the Pacific Ocean basin?

Use the Measure tool to determine the greatest width of new ocean floor created in the Pacific Ocean basin over the last 10 million years. (Caution: See Figure 4 at left.)

24. How much new ocean floor (total width) has been created at this location in the Pacific Ocean basin in the past 10 million years?

25. Use this measurement to calculate the rate of ocean-floor spreading at this location in the Pacific Ocean basin, in centimeters per year. You may need to look back in the left margin of the previous page to review the procedure for converting units.

26. In which ocean basin did you measure the highest spreading rate—the Pacific or the Atlantic?

Quit ArcMap and do not save changes.

Big ideas and questions about forming ocean basins

Three major ideas presented in this activity are still unresolved:

- The oldest ocean-floor rocks are 190 Ma, whereas the oldest continental rocks are 3.8 billion (3800 million) years old. Yet we know from fossil and rock records that the oceans existed 3.8 billion years ago. Has old ocean-floor rock disappeared? Was it converted into something else, like continental rock? Have the oceans been growing at the expense of the continents?
- Ocean-floor rock ages increase laterally with distance away from the youngest rocks. The youngest rocks are typically (though not always) found near the middle of an ocean, and the oldest rocks are typically found near the continents. This implies a mechanism whereby ocean crust is made at these central areas, but does not explain how.

- New ocean floor seems to form at different rates in different ocean basins and over time. What is the impact of one ocean basin generating new rock more rapidly than another?

Keep these ideas and questions in mind as you explore the ocean basins in greater detail later in this unit.

Reading 1.3

Ocean origins

How did the oceans form?

Scientists believe that the oceans developed early in Earth's history—at least 3.8 billion years ago. When our planet formed 4.6 billion years ago, heat from compression, nuclear reactions, and collisions with solar system debris caused the early Earth to melt. Molten materials separate, or differentiate, into layers based on **density**. In Earth, this process formed concentric layers, beginning with a dense iron-nickel core, surrounded by a thick mantle made of magnesium, iron, silicon, and calcium (Figure 1 at left). The rock with the lowest density rose to the surface to form a thin crust. Together, the crust and rigid part of the upper mantle form the **lithosphere**, with an average thickness of around 100 km. Beneath the lithosphere is the **asthenosphere**, a ductile or **plastic** region of the upper mantle.

During differentiation, **volatile** elements trapped within the early Earth rose towards the surface and were vented by volcanoes to form our atmosphere. This early atmosphere consisted mainly of hydrogen and helium, very light elements that were easily lost to space. Around 3.9 billion years ago, as Earth condensed further, a second atmosphere formed, but it contained only traces of free oxygen (O_2). This early atmosphere reflected much of the solar radiation striking Earth, allowing the surface to cool and water vapor to condense into rain. At first, Earth's surface was too hot for liquid water to exist on the surface. Eventually, it cooled enough for water to accumulate, forming the early oceans. Scientists think that this is when the crust began to differentiate into two types—continental and oceanic crust—because the process that forms granite, the most common type of continental rock, requires the presence of water.

The earliest forms of life—blue-green algae—developed in the oceans, where water offered protection from the harmful ultraviolet (UV) radiation that penetrated Earth's early atmosphere. Through the process of photosynthesis, these organisms produced their own food using carbon dioxide, water, and energy from the sun, and released an important by-product—oxygen.

1. Why were there no oceans during the first 800 million years of Earth's history?

2. Why do scientists think that the continental rocks could not have formed before the oceans?

density — the ratio of mass to volume of an object or substance. Mathematically,

$$\text{density} = \frac{\text{mass}}{\text{volume}}$$

plastic (adjective) — able to change shape without breaking.

volatile (adjective) — easily evaporated.

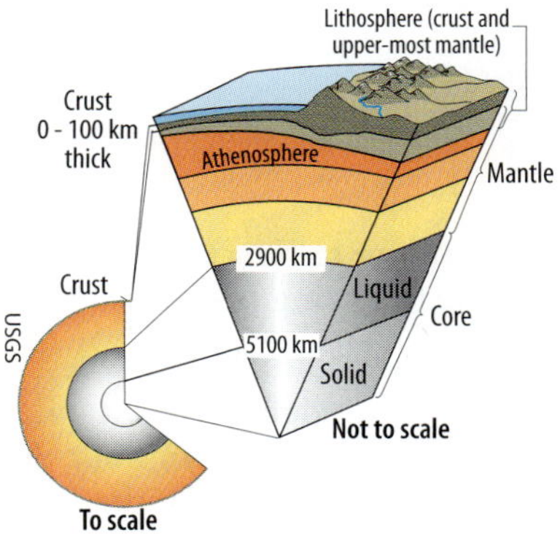

Figure 1. Earth's internal structure is defined by differences in density, composition, and the physical state of the material.

Why are the ocean basins so much younger than the continents?

To understand the answer to this question, you must examine the processes that form and modify oceanic and continental lithosphere.

USGS

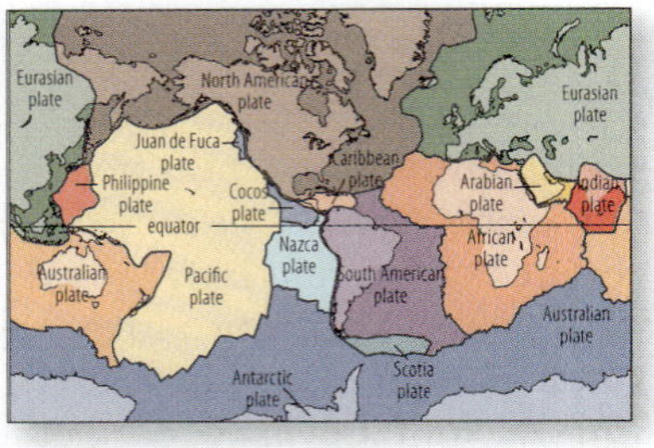

Figure 2. Earth's major lithospheric plates.

USGS

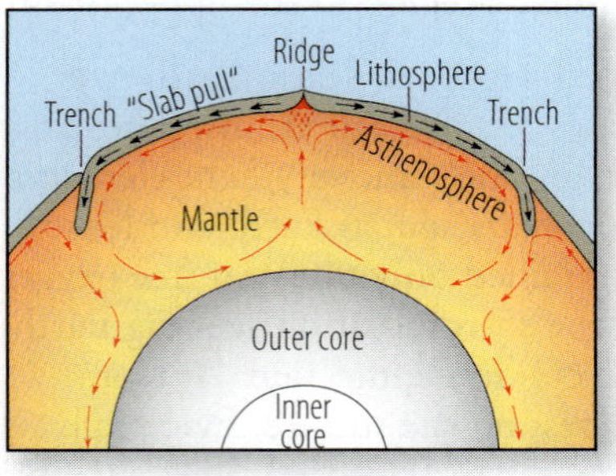

Figure 3. Convection in Earth's mantle tranports heat energy to the surface.

Plate tectonics

Earth's outermost layer, the **lithosphere**, is made up of the crust and uppermost part of the mantle. The theory of **plate tectonics** states that the lithosphere is fragmented into 20 or so rigid plates that are moving relative to one another (Figure 2 at left). The plates move atop the more mobile, plastic asthenosphere.

Earth's core reaches temperatures over 6000 °C. About half of this heat comes from the decay of naturally occurring radioactive minerals. The rest is left over from the heat generated when Earth formed. Heat energy travels slowly from Earth's interior to the surface by the processes of **conduction** and **convection**.

In conduction, heat energy moves by collisions between molecules. Convection occurs as rock heats up, expands, becomes less dense, and rises toward the surface (Figure 3 at left). Near the surface it cools, contracts, becomes denser, and slowly sinks deep into the mantle again. Convection cools Earth more efficiently than does conduction. Still, it takes a long time—billions of years—to cool an object the size of Earth. As long as sufficient heat reaches the upper mantle, Earth's lithospheric plates will continue to move.

Types of plate boundaries

There are three types of plate boundaries, classified according to how adjacent plates are moving relative to each other (Figure 4 on the following page).

At **divergent boundaries**, plates move away from each other, forming rift valleys on land and spreading ridges in the oceans. These mid-ocean ridges form the longest mountain belt on Earth, extending over 60,000 km on the ocean floor. At the ridge, molten rock or magma wells up between separating plates and intrudes into the plates, solidifying into new oceanic crust along the ridge crest. Pressure from beneath the ridge, combined with gravity, causes the plates to continually break during earthquakes and spread apart, allowing more magma to rise to the surface, creating more new crust. As the new crust moves away from the ridge, it cools and contracts, becoming denser. Over time, the upper mantle beneath this crust cools, thickening the lithosphere. As the density of the crust and lithosphere increase, they sink deeper into the mantle and the ocean basin deepens. The high ridge at the plate boundary is stationary while the plates themselves move away, revealing their generally flat topography.

At **convergent boundaries**, the plates are moving toward each other (Figure 4). If the two plates are topped with continental crust, they will collide, causing the crust to pile up and create mountains. If one or both plates are composed of only oceanic lithosphere, the older denser oceanic plate will plunge down into the mantle, in a process called **subduction**. Both plates are bent downward along their line of contact, forming a deep trench at the surface along the plate boundary.

The subducting oceanic plate carries with it sediments and water from the surface. Near a depth of 150 km, the descending plate heats up sufficiently for some of the material to melt. The water and sediments change the chemistry of the **magma**

Plate boundary animations

To see animations and visuals of the plate boundaries, launch ArcMap, locate and open the etoe_unit_1.mxd project file, then open any data frame. Click the Media Viewer button and choose any of the following movies or visuals from the media list.

- **Ridge Spreading and Subduction**
- **Transform Boundary**
- **Ocean-Continent Convergent Boundary**
- **Continent-Continent Convergent Boundary**
- **Divergent Boundary**

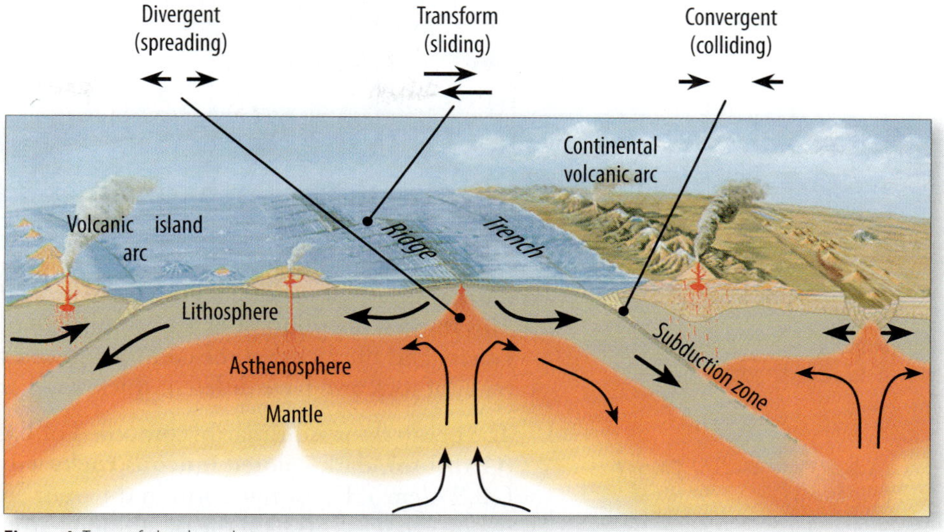

Figure 4. Types of plate boundaries.

(molten rock within the Earth), that rises toward the surface and cools to form new continental rock.

At **transform boundaries**, plates move past each other (Figure 4). The most famous transform boundary in North America is the San Andreas fault in California. Smaller transform faults accommodate movement along spreading ridges, allowing segments of the ridge to expand independently. These types of boundaries and faults are characterized by frequent earthquakes. Transform boundaries are important in rearranging the plates, but are not directly involved in creating or recycling crustal rock.

Tectonic plates have been created, destroyed, and rearranged countless times over Earth's 4.6 billion-year history. New ocean floor forms at mid-ocean ridges, and old ocean floor is recycled into the mantle at trenches. Gravity will not allow Earth to expand or shrink, so the net amount of new rock formed must balance the amount being recycled. Because the oldest oceanic crust can be dated to about 190 Ma, it is reasonable to say that the entire ocean floor, covering some 70 percent of Earth's surface, is completely recycled every 190 million years at the current rates of motion.

weathering — the physical disintegration and chemical altering of rocks and minerals at or near Earth's surface.

erosion — the process by which Earth materials are worn away or removed from their source area by wind, water, or glacial ice.

On the other hand, the low-density continental rock remains on the surface and is not recycled except through **weathering** and **erosion**. In fact, continental rocks that formed 3.8 billion years ago can still be found on Earth's surface today, along with continental rock that formed more recently. Thus, plate tectonics and density differences between continental and oceanic rocks explain why the ocean basins are much younger than the continents.

3. If new ocean crust is formed at mid-ocean ridges, where would you expect the oldest ocean floor to be found?

4. At which type of plate boundary does each of the following types of crust form?
   a. Oceanic —
   b. Continental —

Why are the ocean basins so much lower in elevation than the continents?

Earth's surface has two main features: deep ocean basins and elevated continents. This uneven distribution of topography is no coincidence—it occurs because of gravity and the fact that the continental and oceanic crusts are made of different types of rock and have different thicknesses and densities.

igneous rock — rock formed by the cooling and crystallization of magma or lava.

The ocean floor is primarily composed of **basalt** (buh-SALT) and **gabbro** (GAB-bro), dark-colored **igneous rocks** with the same composition and density (~2.9 g/cm³), but which form in different environments. Basalt comes from magma that erupts *onto* the ocean floor, whereas gabbro crystallizes from magma that cools *within* the oceanic crust. Oceanic crust averages around 6 km thick, except at oceanic plateaus, where it may reach a thickness of up to 40 km.

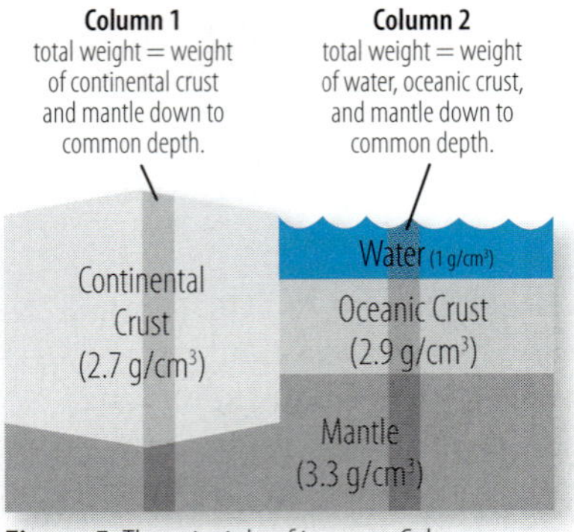

Figure 5. The principle of isostasy. Columns with the same area and to the same depth have equal weight.

The continents are composed primarily of **granite**, a light-colored igneous rock with a density of around 2.7 g/cm³. Continental crust averages about 40 km thick, varying from as little as 20 km to as much as 70 km at high-elevation continental plateaus such as the Tibetan Plateau (the Himalayas).

The existence of deep ocean basins and high-elevation continents is due to the principle of **isostasy** (eye-SAHS-tuh-see), which states that *the total weight of any column of rock and water from Earth's surface to a constant depth is approximately the same as the weight of any other column of equal area from its surface to the same depth* (Figure 5 at left). Differences in the heights of the columns are due to differences in the density and thickness of the materials in the columns. The **equilibrium** between columns is maintained by the plastic flow of material within Earth's mantle. Though rock in the mantle is generally not hot enough to melt, it is warm enough to flow or change shape very slowly when subjected to a force, much like modeling clay or Silly Putty®.

equilibrium — state of balance; condition where no overall changes occur in a system.

In the simplest example, isostasy is the principle observed by Archimedes (ahr-kuh-MEED-eez) in the third century B.C., when he noticed that an object immersed in water displaces a volume of water equal to the volume of the object. When two columns of different depth and weight form on Earth's surface, the mantle will slowly flow out from beneath the added load. The region with greater weight will have less mantle beneath it than the region with lesser weight.

If crust or water is removed from the top of the column, mantle material flows in underneath the column to compensate for the reduced weight, pushing the column back up slightly. On the other hand, if more crust or water is piled atop the column, mantle material flows away from the bottom of the column to equalize the weight, causing the column to sink back down slightly. The mantle behaves in a manner similar to that of water when an ice cube is placed in a container of water. Water beneath the ice flows outward and upward and the ice sinks downward until the water and ice reach equilibrium.

Exploring isostasy

To explore the principle of isostasy, point your Web browser to:

atlas.geo.cornell.edu/education/student/isostasy.html

You can also:

- Open the **etoe_unit_1.mxd** project file.
- Open any data frame.
- Click the Media Viewer button.
- Choose **Explore Isostasy.**

5. If a 1-km thick layer of material were removed from the top of a continent by erosion, would the elevation of the continent decrease by 1 km? Explain.

USGS and University of California, Berkeley

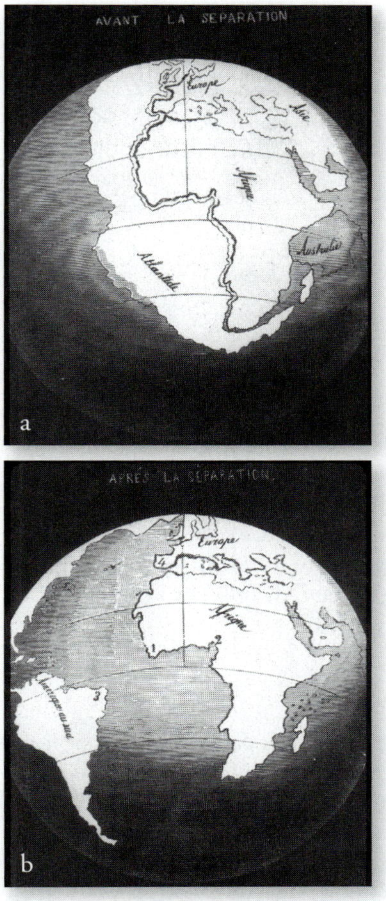

Figure 6. In 1858, geographer Antonio Snider-Pellegrini noted how the American and African continents may once have fit together (a), then later separated (b). In the early 1900s, German climatologist Alfred Wegener expanded on this idea in his theory of *continental drift*.

How do we know what the continents and oceans looked like in the past?

Over the past 200 years, scientists have collected evidence of Earth's changing surface through mapping and monitoring topography; studying earthquakes, volcanoes, and fossils; and measuring the ages of rocks.

As early as the 1600s, cartographers noticed that the coastlines of continents sometimes appeared to match up, even across oceans, like pieces of a puzzle (Figure 6 at left). Then, in the 1800s, scientists discovered that continents separated by vast oceans contained similar landforms, rocks, and fossils. This discovery suggested that today's continents are in very different places than they were in the past (Figure 7 below). The theory that the continents moved over time to their present locations is called **continental drift**. None of these observations provided an explanation for how entire continents could move over Earth's surface, so the theory was not widely accepted. The key evidence for the mechanism driving plate tectonics came in the mid-1900s, through detailed studies of earthquakes, volcanoes, and the ocean floor.

USGS

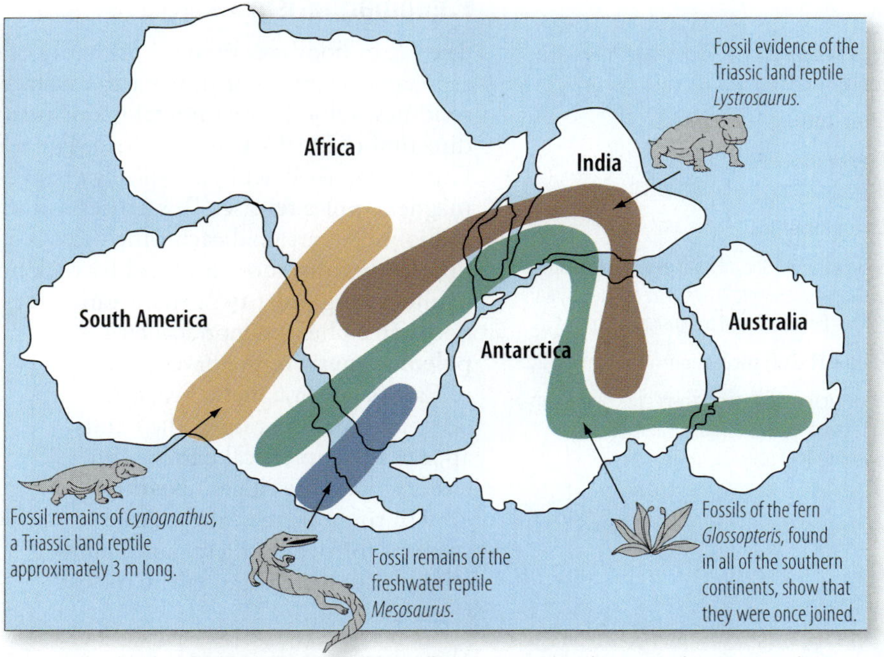

Figure 7. Fossil evidence points to a time around 250 million years ago when the present-day continents shown above were joined in a larger continent, now called *Gondwanaland*.

How old is that rock?

The ages of rocks are determined through a variety of methods. Some techniques allow us to rank the ages of rocks relative to one another, whereas others allow us to assign numerical ages to individual rocks.

Relative dating

Relative dating methods do not tell us the actual ages of rocks, but they do allow us to determine the **chronological** (krah-nuh-LAH-ji-kuhl) sequence in which rocks formed. Using a series of **stratigraphic principles** (see sidebar), scientists can often reconstruct the history of a sequence of rock layers (**strata**) and geological events. Relative dating techniques are most useful on the continents, where rocks have been uplifted and exposed by weathering and erosion, or can be drilled into relatively easily.

chronological — arranged in order of time of occurrence.

Stratigraphic principles

Original horizontality

Sedimentary layers and lava flows are originally deposited as relatively horizontal sheets called *strata* (singular: stratum).

Lateral continuity

Lava flows and sedimentary layers extend laterally in all directions until they thin to nothing or reach the edge of the basin of deposition.

Superposition

In undisturbed rock layers or lava flows, the oldest is at the bottom and the youngest is at the top.

Inclusions

A piece of rock included in another rock or layer must be older than the rock or layer in which it has been incorporated.

Cross-cutting

A feature that cuts across another feature must be younger than the feature that it cuts.

Unconformities

Surfaces called unconformities represent gaps in geologic time where layers were not deposited or have been removed by erosion.

Faunal (fossil) succession

Plants and animals evolve over time, and each time period can be identified by a unique assemblage of plant and animal fossils.

Absolute dating

Absolute dating techniques use laboratory analysis to determine how much time has passed since the rocks formed. Most igneous rocks — rocks that form from cooled magma or lava — contain tiny amounts of radioactive elements. These radioactive **parent** elements break down into different **daughter** elements over time at a known rate. By precisely measuring and comparing the amount of parent to daughter elements in a rock, a process called **radiometric dating**, scientists can determine how many years ago the rock formed. Absolute dating techniques are useful for both oceanic and continental rocks, but relatively few oceanic rocks have been dated radiometrically because obtaining good samples is difficult and costly. In practice, absolute and relative dating techniques are often used together to determine approximate ages of rocks.

Paleomagnetism

The ocean floor is composed primarily of an igneous rock called basalt that forms as magma erupts at or near Earth's surface and cools rapidly. Before the magma solidifies, some of the mineral grains turn like tiny compasses, preserving the direction of Earth's magnetic field. For reasons that are not fully understood, Earth's magnetic field periodically changes polarity. That is, the north and south magnetic poles reverse. This process occurs irregularly every tens of thousands to millions of years; and each time there is a reversal, minerals in the new rock align differently from those in the older rock next to it. Thus, reversals appear as bands of alternating polarity in the ocean floor. This phenomenon, called **paleomagnetism**, preserves a record of Earth's past magnetic fields. At oceanic spreading ridges, the bands appear as symmetrical patterns on either side of the ridge. Like the growth rings of trees, the unique patterns formed by these bands allow scientists to assign ages to the rocks (Figures 8 and 9).

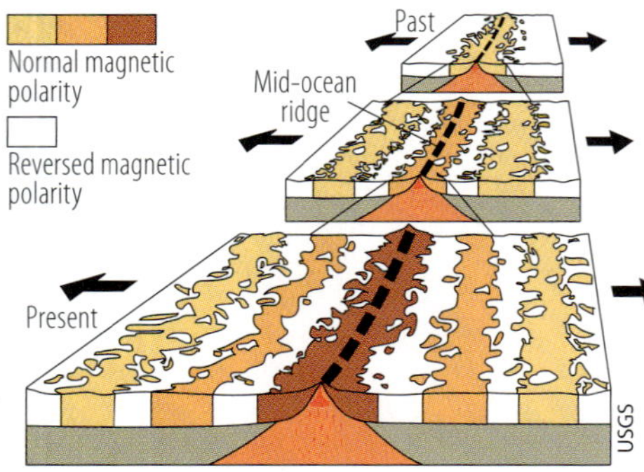

Figure 8. As plates move apart, symmetrical patterns of normal and reversed polarity are preserved in the rocks.

Ships crisscross the oceans, recording these patterns with sensitive **magnetometers** — devices that

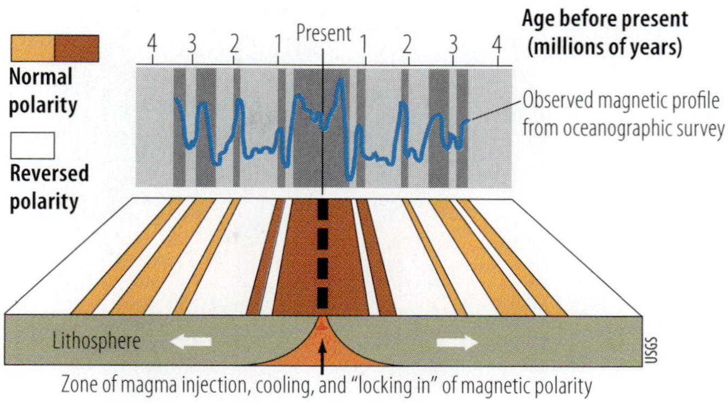

Figure 9. Magnetic polarity profiles are analyzed to determine the age of the ocean floor.

measure the strength and direction of the magnetic field preserved in oceanic rocks. These data confirm that new ocean floor forms at ridges and moves away from the ridge, providing additional confirmation of plate tectonic theory. In addition, researchers have used paleomagnetic data to measure the rate of ocean-floor spreading and to reconstruct the history of ocean basin development.

6. Briefly explain the advantages and disadvantages of the following dating techniques for dating ocean-floor rock.
   a. Radiometric dating

   b. Paleomagnetic dating

In this reading, you have explored how heat and gravity drive the processes that are continuously changing Earth's surface. In the next two investigations, you will examine the ocean floor at increasing levels of detail to look for features that provide evidence of these processes.

Investigation 1.4

Beneath the waves

For most of human history, the ocean's surface has been a forbidding boundary, separating the known from the unknown. Except for the tiny amount of the ocean floor visible in shallow water, people had no idea what lay beneath the waves.

bathymetry (buh-THIH-muh-tree) — measure of water depth in basins such as lakes, oceans, and rivers. *Bathy-* comes from a Greek word for *depth,* and *-metry* comes from another Greek word meaning *to measure.*

Today, we have the ability to gather detailed information about the age, composition, and other characteristics of the ocean floor. This knowledge is critical for understanding the processes that shape Earth's surface. In this activity, you will investigate the *bathymetry* of the ocean basins to more fully understand the features of the ocean floor and the processes that shape them.

The five ocean basins

Over 70 percent of Earth's surface is covered by a single, interconnected body of water that is somewhat arbitrarily divided into five basins — the Arctic, Atlantic, Indian, Pacific, and Southern Oceans. Next, you will examine the location, size, and depth of each ocean basin.

- Launch ArcMap, and locate and open the **etoe_unit_1.mxd** file.

Refer to the tear-out Quick Reference Sheet located in the Introduction to this module for GIS definitions and instructions on how to perform tasks.

- In the Table of Contents, right-click the **Ocean-Floor Topography** data frame and choose Activate.
- Expand the **Ocean-Floor Topography** data frame.

This data frame shows the ocean basins, each outlined in a different color. If you find the outlines hard to see, turn off the **Countries** layer.

- Click the Identify tool.
- In the Identify Results window, select the **Ocean Basins** layer from the drop-down menu.
- Next, click within each ocean basin and read the name of the basin in the Identify Results window.

1. Label the five global ocean basins on Map 1 on the next page.

- Close the Identify Results window.

NOAA

Figure 1. Manual depth sounding with a weighted cable. It was difficult to obtain accurate depth measurements this way. Sometimes, it was hard to determine the exact location of the ship, to ensure that the line dropped straight down, and to know when the weight hit the bottom.

Mapping the ocean floor

Early depth measurements were often labor-intensive and prone to error (Figure 1 at left). Fortunately, today we can map the features of the ocean floor on a global scale using satellites. The satellites measure tiny variations in the height of the ocean surface caused by gravity, which correspond directly to the depth of the ocean floor.

Are there two Pacific Ocean basins?

No. Earth's curved surface has to be "split" somewhere in order to make a flat map. Here, the split was made at 180° longitude, dividing the Pacific Ocean basin into two parts. One part appears on the far right and the other on the far left of the map.

Map 1 — Global ocean basins

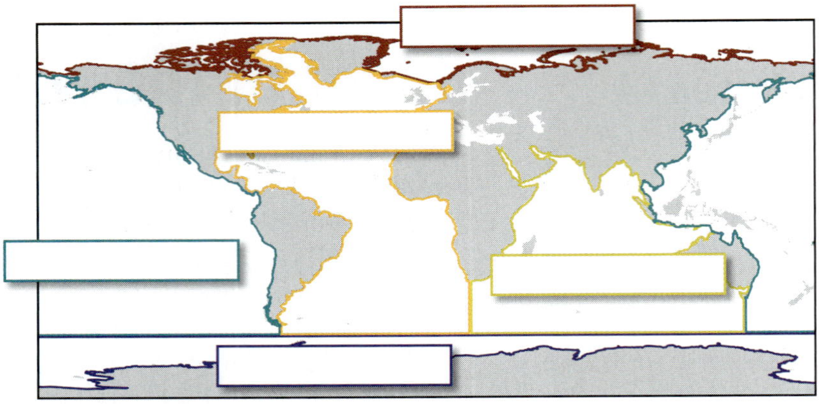

Bathymetric profiles

One way to examine ocean depth data is by creating bathymetric profiles that show what the ocean floor would look like if you sliced through it and viewed it from the side. These profiles illustrate the shape of the basin and reveal submerged features. Next you will create increasingly detailed profiles to see how our understanding of the ocean basins has improved over the past 150 years.

Like early sailors and explorers, your first profile of the Atlantic Ocean Basin will be based on only a few depth measurements.

- Turn on the **Atlantic Crossing** layer.

This layer displays the path of a ship crossing the Atlantic Ocean from Florida to Africa.

- Click the QuickLoad button.
- Select **Spatial Bookmarks**, choose **Atlantic Crossing**, and click **OK.**

Simple profiles

A profile through a bathtub would look something like this:

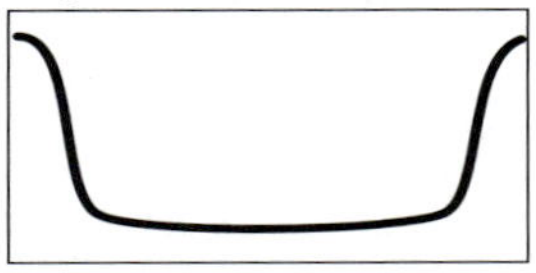

Whereas a profile of a mountain might look like this:

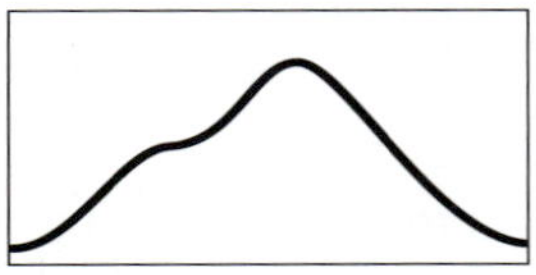

2. On Graph 1, sketch what you *think* the profile (an outline) of the ocean floor across the Atlantic Ocean basin will look like along the ship's path from North America to Africa. (See sidebar for examples.)

Graph 1 — Predicted depth profile of Atlantic Ocean Basin

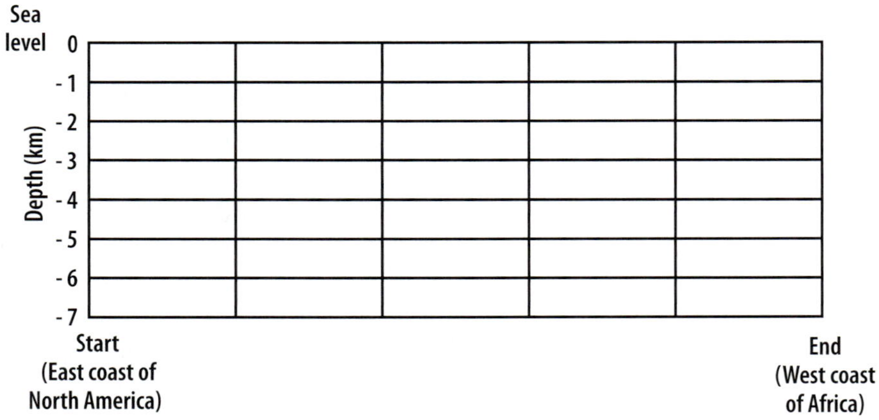

3. Explain your reasoning for the shape of the profile you drew. How did you decide where to draw the shallow and deep parts of the ocean?

Turn off the **Ocean Basins** layer.

Turn on the **Atlantic Bathymetry** layer.

Turn on the **Depth Measurements #1** layer.

The **Depth Measurements #1** layer displays four green points, labeled 1 through 4, at intervals along the ship's path. Each point represents a location where the ship stopped to measure and record the ocean depth.

Click the Identify tool.

In the Identify results window, select the **Depth Measurements #1** layer from the drop-down menu.

Next, click on each of the four points along the path and read the depth at each point from the Identify Results window.

4. Record the depth values for each of the four points in Table 1. Round values to the nearest 0.1 km.

Table 1 — Atlantic Ocean Basin depth measurements #1

Point	Depth (km)
0	*0*
1	
2	
3	
4	
5	*0*

5. Plot the depth values from Table 1 on Graph 2. Draw a smooth line through each point beginning at point 0 (sea level) on the North American coast and ending at point 5 (sea level) on the African coast.

Graph 2 — Atlantic Ocean Basin depth profile 1

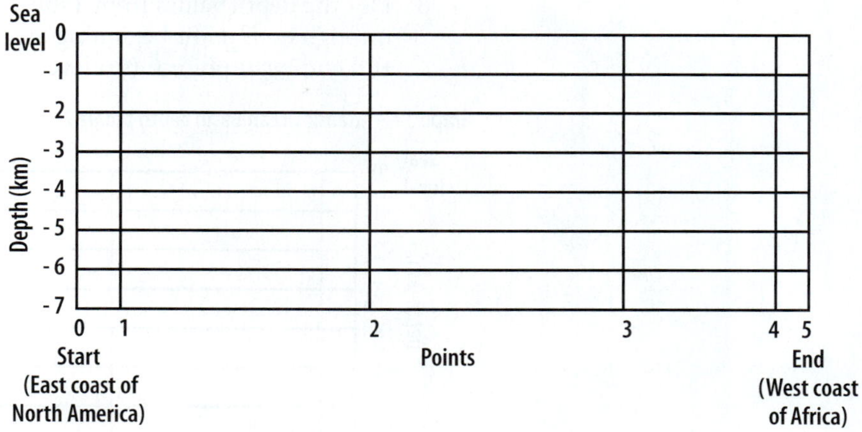

6. Describe the shape of the Atlantic Ocean basin as it appears in Graph 2. How does it differ from your predicted profile in Graph 1?

Why is the depth negative?

The elevation of a mountain like Mount Everest is based on measuring from sea level to the top of the mountain. In these materials, we will use negative values of elevation to express the depth of a body of water from sea level downward. Therefore, we will say the depth of the Mariana Trench is – 11,035 meters.

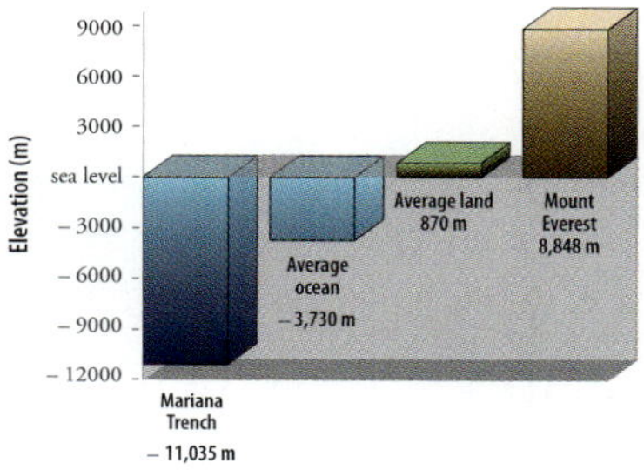

Close the Identify Results window.

Next, you will create a bathymetric profile of the Atlantic Ocean using measurements taken at different points along the ship's path.

Turn off the **Depth Measurements #1** layer.

Turn on the **Depth Measurements #2** layer.

This layer displays five blue points that represent the locations of the new depth measurements.

Click the Identify tool.

In the Identify results window, select the **Depth Measurements #2** layer from the drop-down menu.

Next, click on each of the five points along the path and read the depth at each point from the Identify Results window.

7. Record the depth values for each of the five points in Table 2. Round values to the nearest 0.1 km.

Table 2 — Atlantic Ocean Basin depth measurements #2

Point	Depth (km)
0	0
1	
2	
3	
4	
5	
6	0

Close the Identify Results window.

8. Plot the depth values from Table 2 on Graph 3. Draw a smooth line through each point beginning at point 0 (sea level) on the Florida coast and ending at point 6 (sea level) on the African coast.

Graph 3 — Atlantic Ocean Basin depth profile 2

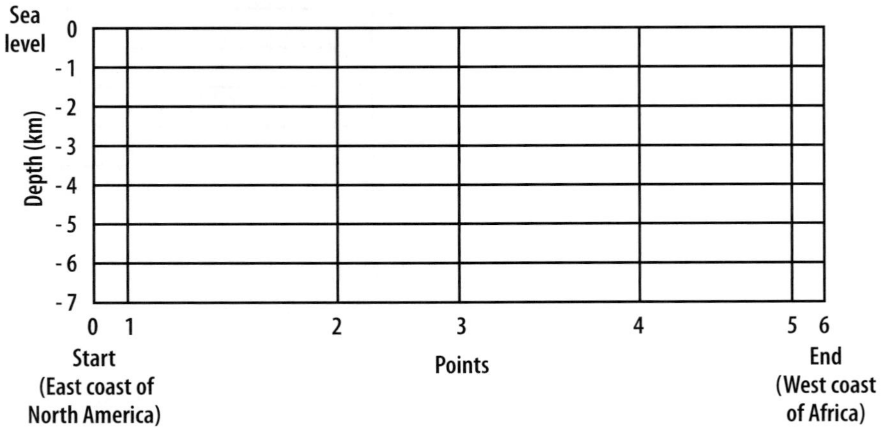

9. Compare Graph 3 to Graph 2. Describe any "new" features that appeared in Graph 3 that are not visible in Graph 2.

10. How do the number and selection of measurement locations affect the profile?

NOAA

Figure 2. Sonar measures the depth of the water by determining the time it takes sound waves to travel to the ocean floor and back.

- Turn off the **Depth Measurements #2** layer.
- Turn off the **Atlantic Crossing** layer.

In the last two profiles you created, you used depth measurements taken at only a few locations. Next, you will create a bathymetric profile of the Atlantic Ocean basin along the ship's path using nearly continuous depth measurements like those generated by sonar (Figure 2).

- Click the Select By Location button.
- In the Select By Location window, construct the query statement:

  I want to **select features from** the **Atlantic Bathymetry** layer that **intersect** the features in the **Atlantic Crossing** layer.

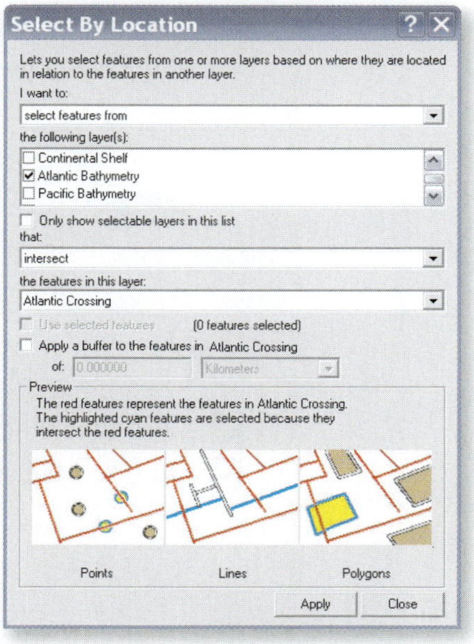

- Click **Apply**.
- Close the Select By Location window.

Sixty bathymetry measurements along the **Atlantic Crossing** path should be highlighted. Next, you will construct a graph using these depth measurements.

- Select the **Atlantic Bathymetry** layer.
- Click the Open Attribute Table button to open the **Atlantic Bathymetry** attribute table. (Warning: Do not click anywhere in the table window, or you will lose the features you selected in the previous step.)
- Click **Options** and choose **Create Graph**.
- Under Graph type, click **Column**. Do not alter the Graph subtype. Click **Next**.
- In the next window, select the **Atlantic Bathymetry** layer and the **Depth (km)** field. Be sure to check the box to use the selected set of features or records. Click **Next**.
- In the next window, click **Finish**.
- Close the **Atlantic Bathymetry** Attribute Table to view the graph.

The graph now displays depth measurements along the ship's path across the Atlantic Ocean, creating a bathymetric profile.

11. Using this graph, describe the features that characterize the floor of the Atlantic Ocean basin. (Hint: Click and drag on the corner of your graph to enlarge it. This will help you see the ocean floor in more detail.)

12. How did increasing the number of depth measurements change your view of the ocean floor bathymetry in the Atlantic Ocean basin?

- Close the graph window.
- Click the Full Extent button to view the entire map.
- Quit ArcMap and do not save changes.

Investigation 1.5

Ocean basin features

Early explorers learned (often the hard way) that the ocean floor is *not* smooth like a bathtub. But how does bathymetry vary among ocean basins or within a basin? And how is the shape of the basin related to the age of the rocks or the way the basin formed? To answer these questions, you will begin by looking at the general characteristics of each basin. Then you will examine, in detail, the floors of the ocean basins.

Launch ArcMap, locate and open the **etoe_unit_1.mxd** file.

Refer to the tear-out Quick Reference Sheet located in the Introduction to this module for GIS definitions and instructions on how to perform tasks.

In the Table of Contents, right-click the **Ocean-Floor Topography** data frame and choose Activate.

Expand the **Ocean-Floor Topography** data frame.

Turn on the **Ocean Basins** layer.

Click the Identify tool.

In the Identify Results window, select the **Ocean Basins** layer from the drop-down menu.

Next, click within each ocean basin to obtain its average depth and surface area.

1. Record the average depth and surface area of each ocean basin in Table 1. Round all values to the nearest 0.1 km.

Table 1 — Global ocean basin statistics

Ocean	Average depth *km*	Depth rank	Surface area *million* km^2	Area rank
Arctic				
Atlantic				
Indian				
Pacific				
Southern				

2. Rank the oceans by the amount of surface area and by depth using the numbers 1 through 5, with 1 being the largest or deepest and 5 being smallest or shallowest.

3. Which ocean basin is the shallowest? How much shallower is this ocean basin than the next shallowest ocean basin?

Close the Identify Results window.

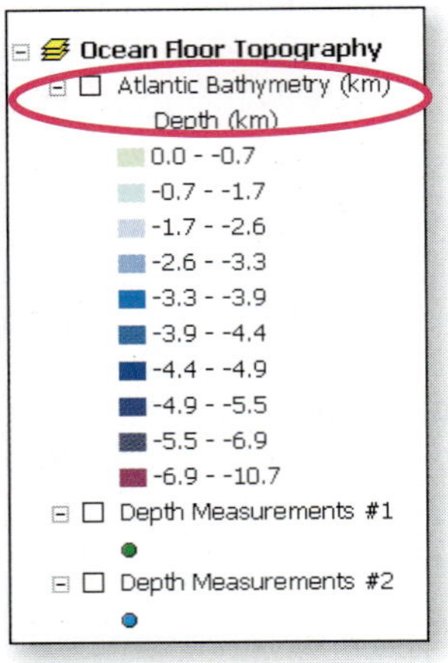

Figure 1. The new Atlantic Bathymetry layer produced by the quickload process will be located at the top of the Table of Contents.

Clearly, the Arctic Ocean is much shallower than the other oceans. Next, you will explore why.

- Turn on and expand all bathymetry layers: **Atlantic Bathymetry**, **Arctic Bathymetry**, **Indian Bathymetry**, **Pacific Bathymetry**, and **Southern Bathymetry**.

Unlike the other layers, the depth of the Atlantic Ocean is currently represented by a single shade of blue.

- Right-click the **Atlantic Bathymetry** layer and choose Remove.

You will now add a new **Atlantic Bathymetry** layer with the depth classified like the other bathymetry layers.

- Click the QuickLoad button.
- Select **Layers**, choose **Atlantic Bathymetry**, and click **OK**.
- Locate the new **Atlantic Bathymetry** layer in the Table of Contents (Figure 1).
- Click on the **Atlantic Bathymetry** layer and drag it down the Table of Contents so that it sits just above the **Pacific Bathymetry** layer.
- Turn on the **Atlantic Bathymetry** layer.

4. Using the Bathymetry legends, list the color(s) and depth range of
   a. the shallowest part of the ocean.

   b. the deepest part of the ocean.

- Turn on and expand the **Continental Shelf** layer.

This layer shows where continental rocks are submerged beneath ocean waters. This submerged land, called the *continental shelf*, extends as far as several hundred kilometers offshore and to a depth of about 200 m (Figure 2). Beyond the edge of the shelf, the steep *continental slope* drops away toward the *continental rise* and eventually the deep-ocean floor.

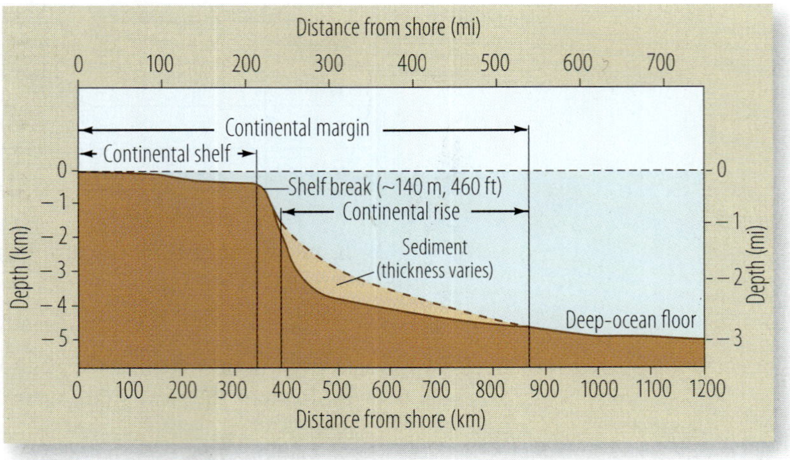

Figure 2. The continental margin.

5. Which ocean has the greatest area of continental shelf?

The existence of the continental shelf is not the only reason for shallow ocean floors. Using depth as a guide, locate the submerged mountain range in the Atlantic Ocean basin.

6. Which of the other ocean basins contain similar submerged mountains?

Turn off the **Continental Shelf** layer.

The submerged mountain range you discovered while examining the Atlantic Ocean bathymetry is the Mid-Atlantic Ridge, an example of a spreading ridge. The discovery of the Mid-Atlantic Ridge in the 1930s caused scientists to wonder about the processes that form this type of underwater ridge system and what other types of features may exist on the ocean floor.

Ridge flyby animation

To view a simulated "flight" along a ridge, click the Media Viewer button and choose **Mid-Ocean Ridge Flyby** from the media list.

NSF/RIDGE

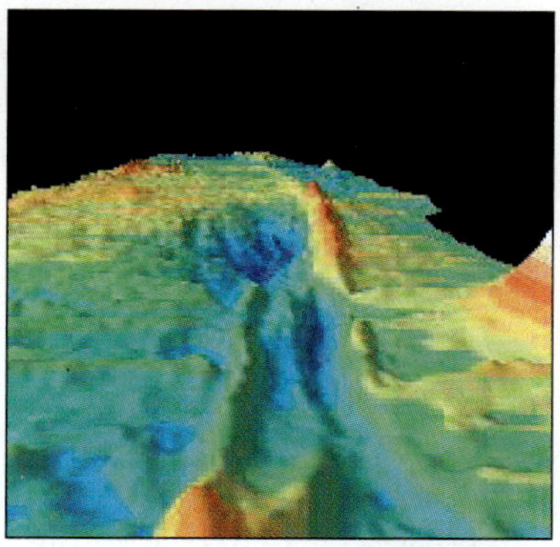

Turn on the **Ridges** layer.

This layer displays the locations of mid-ocean ridges.

7. On Map 1, sketch the locations of the oceanic spreading ridges. Turn off all of the layers but the **Ridges** and **Countries** layers to see the mid-ocean ridges more easily.

Map 1 — Oceanic ridges and trenches

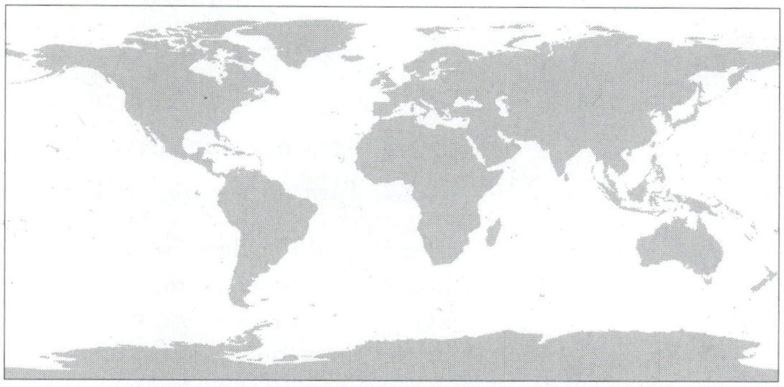

Turn on and expand the **Ocean-Floor Age** layer.

8. Describe any patterns you observe between the location of the ridges and the age of the ocean-floor rocks.

Turn off and collapse the **Ocean-Floor Age** layer.

Next you will determine if there is a relationship between the age of the ocean floor and its depth. You will examine the rocks with ages between 0 and 120 million years.

9. Describe the relationship between ocean-floor age and depth indicated in Figure 3 on the following page.

Why is the depth negative?

The elevation of a mountain like Mount Everest is based on measuring from sea level to the top of the mountain. In these materials, we will express the depth of a body of water from sea level downward, using negative values. Therefore, we will say the depth of the Mariana Trench is – 11,035 meters.

What is Ma?

The Greek prefix *Mega* represents *one million* in the metric system. The Latin word for year is *annum*. As used in Earth science, Ma stands for *Mega annum*, or *millions of years before present.*

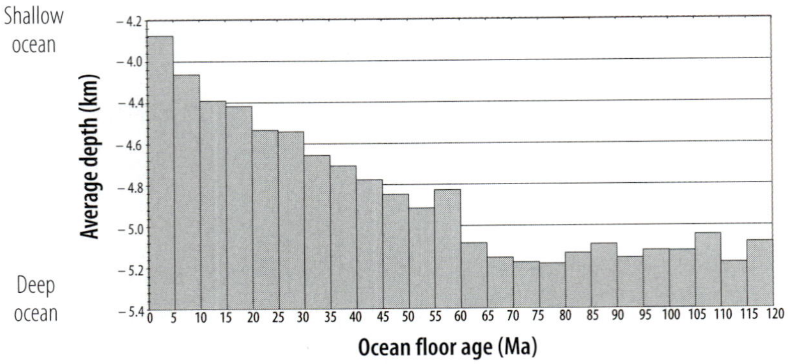

Figure 3. Average depth versus ocean-floor age.

To better understand the processes that form ridges, look at a block diagram of the Mid-Atlantic Ridge.

- Turn on and select the **Block Diagrams** layer.
- Using the Hyperlink tool, click on the profile line that crosses the Mid-Atlantic Ridge to view a block diagram of the ocean floor. (Hint: It is in the same location as the Atlantic Crossing line. Move the Hyperlink tool over the line until a file name appears. Then click to see the image.)

10. Where is the oceanic lithosphere thicker—near the center of the ridge, or far away from it?

- Return to ArcMap and close the block diagram window.
- Turn off the **Block Diagrams** layer.

The spreading ridges are locations where new ocean crust is created. The young, hot, new crust is less dense and rises higher than the surrounding older, colder, denser rock due to *isostasy*—the same process that allows the continents to rise higher than the ocean floor.

In addition to the continental shelves and mid-ocean ridges, there is another type of feature that reveals how the oceans are created and destroyed. You will explore examples of this type of feature next.

Trench flythrough animation

To view a simulated "flight" through a trench, click the Media Viewer button and choose **Trench Flythrough** from the media list.

NSF/RIDGE

Where does old ocean crust go?

If the deepest parts of the ocean basins are not in the middle of the basins, then where are they?

- Turn on all the **Bathymetry** layers.
- Using the color scale of the bathymetry layers, locate three regions anywhere on the ocean floor with depths greater than – 6 km. They should appear dark blue or purple on the map. (Hint: there are many in the Pacific Ocean.)

These regions are *trenches*, the deepest parts of the oceans. At trenches, oceanic lithosphere is subducted and plunges down into the mantle to be recycled (see the **Trench Flythrough** animation in sidebar).

11. On Map 1 (page 35), mark and label (**T**) the locations of three trenches.

- Turn off the **Countries** layer.
- Turn off the **Ridges** layer.

Down the trenches

- Turn on and select the **Ocean Profiles** layer.

This layer displays bathymetric profiles of selected regions of the ocean floor. You may also view an animation to better understand the profiles and how they are related to block diagrams.

- Click the Media Viewer button.
- Choose **Topographic Profiles** from the media list.
- View the movie and close the Media Viewer window.
- Use the Hypelink tool to click on profiles that cross areas that you think contain trenches.

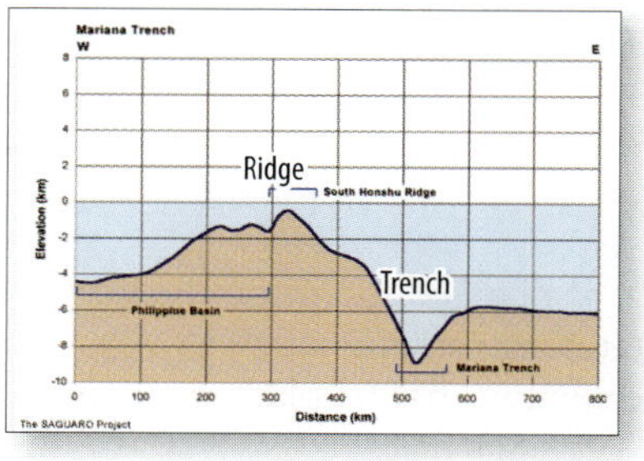

Figure 4. Identifying ridges and trenches in a bathymetric profile.

The deepest points on Earth

The profiles drawn in the **Topographic Profiles** layer do not necessarily cross the deepest part of the trench along its length. The maximum depths (km) recorded for each trech are listed below.

Trench Name	Maximum Depth *km*
Aleutian	8.1
Atacama	8.1
Java	7.7
Kuril	10.5
Mariana	11.0
Middle America	6.6
Puerto Rico	8.6
South Sandwich	8.3

12. Record the name of a trench in the Atlantic Ocean basin. Use the profile graph to estimate the maximum depth along the profile and record it below (see Figure 4 at left for help interpreting bathymetric profiles).

    a. Name —

    b. Maximum Depth —

13. Record the name of any trench in the Pacific Ocean basin. Use the profile graph to estimate the maximum depth along the profile and record it below. Remember that because of the map projection, the Pacific Ocean basin appears on the right- and left-hand sides of the screen.

    a. Name —

    b. Maximum Depth —

- Turn off the **Ocean Profiles** layer.

The process of creating and destroying the ocean floor results in the opening and closing of ocean basins. The formation of ocean floor at ridges often begins by breaking apart a continent. As the ridge develops, so does the ocean floor. Over time, a major ocean basin may form.

You may also view an animation showing how the processes of ridge spreading and subduction work.

- Click the Media Viewer button.
- Choose the **Ridge Spreading and Subduction** movie from the media list.
- View the movie several times and close the Media Viewer window.

What will happen next?

Changes in the shapes of ocean basins occur continuously as new ocean floor is created at ridges and destroyed at trenches. The tectonic plates can reveal where more crust is being created than destroyed.

- Turn on the **Ridges** layer.
- Turn on the **Trenches** layer.

These layers show the locations of spreading ridges and trenches. Each ridge and trench represents a plate boundary.

- Turn off and collapse all bathymetry layers.
- Turn on the **Countries** layer.
- Turn on the **Relative-Plate Motion** layer.

This layer shows pairs of arrows on opposite sides of plate boundaries. The size of the arrow represents the speed at which the plates are moving relative to each other at that location. Each arrow points in the direction that its plate is moving relative to the neighboring plate.

- Use the Zoom in tool to examine the Atlantic Ocean basin, including any ridges, trenches, and relative plate-motion arrows.

14. Would you expect the Atlantic Ocean basin to grow larger or become smaller over time? Explain.

- Click the Full Extent button to view the entire map.
- Next, use the Zoom in tool to examine the Pacific Ocean basin, including any ridges, trenches, and relative plate-motion arrows.

15. Based on the surface features and relative-plate motions, would you expect the Pacific Ocean basin to grow larger or become smaller over time? Explain.

In this investigation, you have seen that discoveries in the past hundred years have led to a revolution in our understanding of the ocean floor and the processes that shape it. This knowledge has provided important clues to help us interpret Earth's present condition, reconstruct its past, and predict its future.

- Quit ArcMap and do not save changes.

Wrap-up 1.6

Why are oceans young?

You have examined the ages of both the ocean floor and the continents. Consider some possible explanations, or hypotheses, that might explain why the oceans are generally so much younger than the continents.

1. Read each of the hypotheses presented below and indicate whether the available scientific evidence suggests each explanation could be true, maybe true, or false. For each hypothesis, describe the evidence that supports your choice.

Hypothesis	True	Maybe True	False	Evidence
H1 — Earth is only 190 million years old.				
H2 — Earth is very old, but the oceans started forming only 190 million years ago.				
H3 — The production of ocean crust has increased over the past 190 million years.				
H4 — Each year some ocean crust is destroyed, so the amount of ocean crust of a given geologic period decreases over time.				

2. With the evidence you have examined, select one of the four hypotheses on the previous page that you feel best explains the young age of the ocean basins. Explain.

Unit 2
Ocean Currents

In this unit, you will

- ***Investigate the forces that drive surface currents in the world's oceans.***
- ***Identify major ocean gyres and their physical properties — temperature, speed, and direction.***
- ***Correlate current direction and speed with global winds.***
- ***Examine ocean salinity and temperature patterns and their relationship to deep-water density currents.***

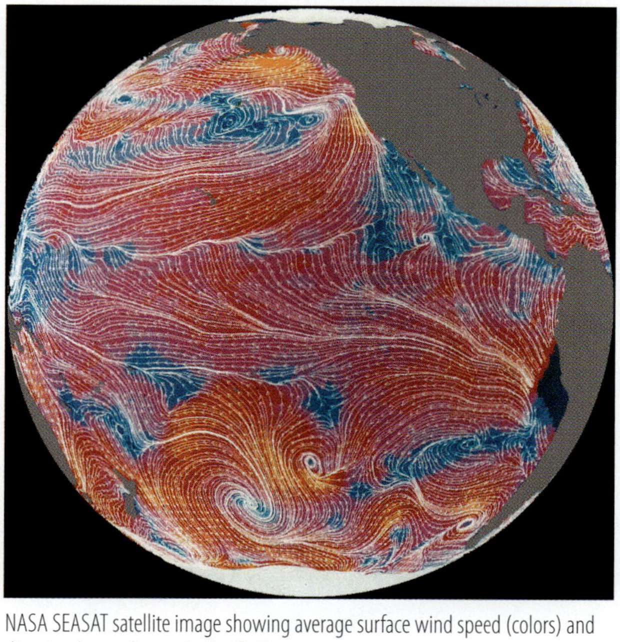

NASA

NASA SEASAT satellite image showing average surface wind speed (colors) and direction (arrows) over the Pacific Ocean.

Warm-up 2.1

A puzzle at 70° N

"Common sense tells us that temperatures increase closer to Earth's equator and decrease closer to the poles. If this is true, the pictures below present a strange puzzle (Figure 1). They show two coastal areas at about the same latitude but on opposite sides of the North Atlantic Ocean. Nansen Fjörd, on the left, is on Greenland's eastern coast, while Tromsø, right, lies on the northwestern coast of Norway. These places are at roughly the same latitude, but their climates could hardly be more different.

Figure 1. A summer day at Nansen Fjörd, on the eastern coast of Greenland (left) and in Tromsø, on the northwestern coast of Norway (right). Both locations are near latitude 70° N.

1. Why do you think the temperatures at the same latitude in Greenland and Norway are so different?

Since the first seafarers began traveling the world's oceans thousands of years ago, navigators have known about currents — "rivers in the ocean" — that flow over long distances along predictable paths.

In 1855, Matthew Maury wrote about the Gulf Stream current, which flows off the east coast of Florida.

> *"There is a river in the ocean. In the severest droughts it never fails, and in the mightiest floods it never overflows; its banks and its bottom are of cold water, while its current is of warm; the Gulf of Mexico is its fountain, and its mouth is the Arctic Sea. It is the Gulf Stream. There is in the world no other such majestic flow of waters."*
>
> —Matthew Maury, *The Physical Geography of the Sea and Its Meteorology*

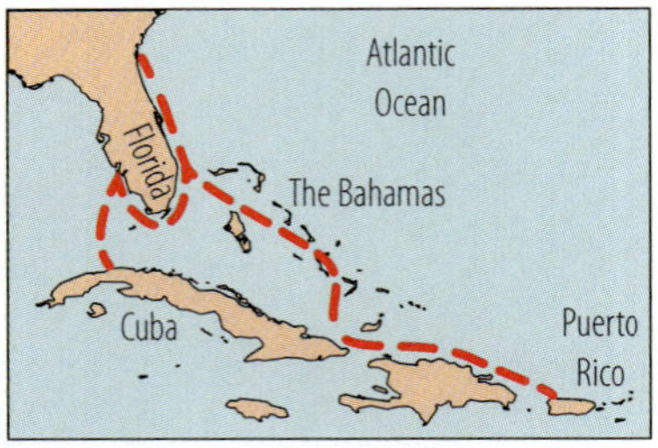

Figure 2. Ponce de León's route.

Maury was not the first person to notice the Gulf Stream. In March 1513, the Spanish explorer Juan Ponce de León left the island of Boriquien (Puerto Rico) in search of the island of Bimini and the legendary Fountain of Youth (Figure 2). Instead, he landed on what is now Florida. After sailing northward along Florida's east coast, he turned around and headed south. While sailing in this direction he discovered that even under full sail with a strong breeze at his back, his ship moved backward in the water! His solution was to maneuver his ship closer to shore and out of the current.

Two hundred fifty years later, Benjamin Franklin, then serving as Deputy Postmaster General, received complaints that ships delivering mail between Boston and England took as long as two months to make the return trip back to America. Merchant ships, which were heavier and took a less direct route than the mail ships, were making the trip back from England in just six weeks.

NOAA

Figure 3. Franklin and Folger's chart of the "gulph stream" current.

With help from his cousin Timothy Folger, a whaling captain, Franklin determined that the returning mail ships were sailing against a strong current that ran along the eastern seaboard and across the Atlantic to the British Isles. Whalers knew about the current, whose plankton-rich margins attract whales, and used or avoided the current as needed to speed their travels. Franklin and Folger offered their chart of the *gulph stream* (Figure 3) to the mail-ship captains, with the promise of cutting their return time in half, but they were largely ignored.

2. What are some factors that might cause ocean water to flow in currents like the Gulf Stream?

3. Explore the idea of what causes ocean currents by comparing how water behaves in a bathtub or small pond, compared to water in the ocean. In Table 1, make a list of differences between the conditions present or acting on a bathtub of water and those in an ocean, and explain how those different conditions might cause currents.

Table 1 — Comparing bathtub water with ocean water

Condition	Bathtub conditions	Ocean conditions	Why this characteristic might cause currents to form
bottom and surface features			
wind			
volume of water			
salinity			
uniformity of temperature			
Coriolis effect (see note at left)			

4. Which of the conditions above do you think are the most important in the formation of ocean currents? Explain.

Coriolis effect

As a fluid like air or water moves over Earth's surface, the planet rotates under it. Relative to the solid Earth, the flow appears to deflect to the right in the Northern Hemisphere and to the left in the Southern Hemisphere. In this way, the Coriolis effect influences the rotation of large-scale weather and ocean-current systems. The Coriolis effect does not influence water in sinks or toilets because the distances involved are very small.

To learn more about the Coriolis effect, point your Web browser to:

http://ww2010.atmos.uiuc.edu/(Gh)/guides/mtr/fw/crls.rxml

NASA/SeaWiFs

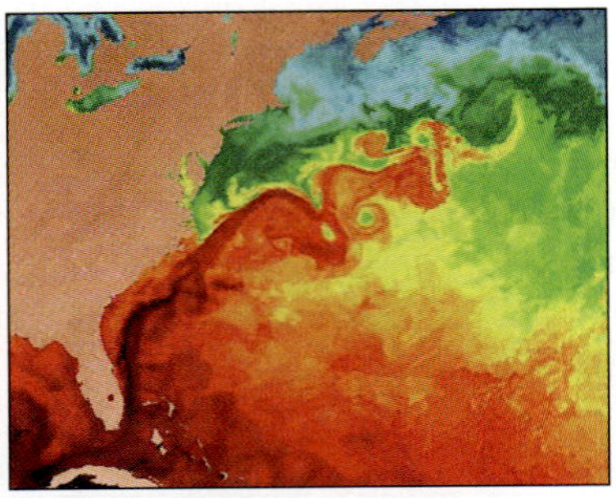

Figure 4. Satellite image of sea-surface temperatures associated with the Gulf Stream off the east coast of North America. Reds and oranges represent warm water, greens and blues cooler water. The warmest water appears dark brown or almost black in this image.

Early nautical charts depicting the Gulf Stream current were useful, but were not entirely accurate. They often assumed that the Gulf Stream began in the Gulf of Mexico when, in fact, it flows westward from the equatorial Atlantic Ocean, turns northward and flows along the East Coast from Florida to the Saint Lawrence Seaway, and then across the Atlantic toward Great Britain (Figure 4).

5. Maury and Franklin both described the Gulf Stream as a warm surface current—that is, its water is warmer than the surrounding ocean. Do you think the ocean also has cold surface currents? Explain your reasoning.

Despite Maury's assertion that "There is in the world no other such majestic flow of waters," the Gulf Stream is not unique. Surface currents have existed in the world's oceans throughout Earth's history, and have influenced life on our planet in important ways.

6. Describe four ways that surface currents might affect you (or another person), either at sea or on land.

   a.

   b.

   c.

   d.

7. Recall the puzzle posed in Question 1 about the extreme climate differences between the coasts of Greenland and Norway. Map 1 shows the location of the Gulf Stream current. On the map, draw the locations of other currents that you think could solve this puzzle. Label each current as warm or cold.

Map 1 — Location of Gulf Stream current

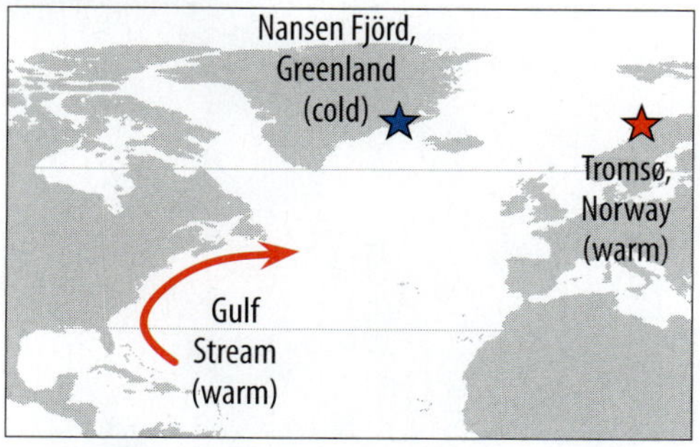

In this unit, you will investigate the forces that drive surface currents and how these currents influence ocean processes and life on Earth.

Investigation 2.2

Oceans in motion

How deep are surface currents?

Surface currents typically extend to depths of less than 400 m.

The oceans are not stagnant, motionless bodies of water. Entire ocean basins gradually change shape, size, and location over millions of years. The oceans are continually active in many other ways as well, across a variety of time scales. Waves rush in and out on an ocean beach within seconds, and sea level rises and falls with the daily tides. Offshore, the movement of the ocean is equally pronounced, with large volumes of water flowing in tremendous currents. Currents can be thought of as vast rivers without banks that transport immense volumes of water around the globe. In this activity, you will explore the characteristics of these surface currents.

Surface currents

To understand how surface currents form, you will begin this investigation by exploring where they are found and the directions in which they flow.

- Launch ArcMap, and locate and open the **etoe_unit_2.mxd** file.

Refer to the tear-out Quick Reference Sheet located in the Introduction to this module for GIS definitions and instructions on how to perform tasks.

- In the Table of Contents, right-click the **Surface Currents** data frame and choose Activate.
- Expand the **Surface Currents** data frame.

This data frame shows the approximate locations and extents of surface currents in each ocean. Different colors represent individual currents, and white represents areas without significant currents.

1. In addition to the Arctic Ocean, there are other large areas where surface currents are absent. In general, in what regions of the oceans are these areas located?

- Turn off and collapse the **Surface Currents** layer.

Current directions

Next, you will search for patterns in the general movement of water within each ocean.

- Turn on the **Current Direction** layer.
- Select the **Current Direction** layer.

In this investigation, current direction is defined as the direction the water is flowing *toward*. Thus, the direction of a current flowing from east to west will be described simply as *west*. The **Current Direction** layer shows the average flow direction of the ocean's surface layer, in one of four cardinal directions (Figure 1 at left): north (N), south (S), east (E), or west (W).

Figure 1. The main, or *cardinal*, directions:north, east, south, and west.

2. Examine the direction of currents in each ocean, then draw circular ocean-current patterns with lines and arrows on Map 1 indicating the direction of motion in each hemisphere and in each ocean. The North Atlantic current line and arrow are already drawn on Map 1 as an example. Remember, there is only one Pacific Ocean; but in this map projection, half of the ocean appears on each side of the map.

Map 1 — Generalized ocean currents

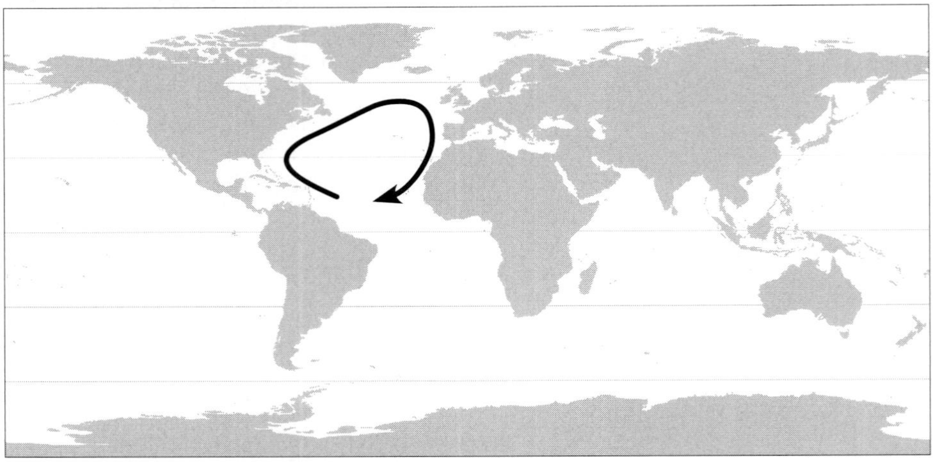

The large, roughly circular paths taken by currents as they flow around the edges of each ocean are called *gyres* (JYE-urz).

3. Examine Map 1 and compare the circulation of gyres in the Northern and Southern Hemisphere. Complete the table below using **CW** (clockwise), **CCW** (counterclockwise), or **B** (both).

Hemisphere	Circulation
Northern	
Southern	

Turn off the **Current Direction** layer.

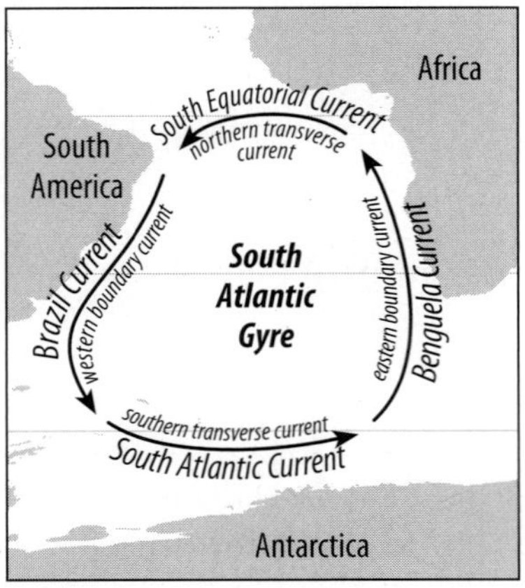

Figure 2. Four types of boundary currents of the South Atlantic gyre.

A closer look at ocean gyres

The four currents that form a gyre flow in a closed circuit around the outer edge of an ocean (Figure 2). The two currents traveling east and west across the ocean are called *transverse currents*, and the two flowing north and south near or along the edges of continents are called *boundary currents*. Next, you will examine the unique temperature and speed characteristics of currents within gyres.

Temperature patterns

In this section, you will characterize global ocean temperatures and investigate how surface currents may influence them.

Turn on the **Water Temperature** layer.

Select the **Water Temperature** layer.

This layer displays the average annual surface temperature of the world's oceans.

Liquid water at -1 °C?

The **Water Temperature** layer shows water as cold as – 1 °C. *Pure* water freezes at 0 °C, but dissolved salts in ocean water lower its freezing point. The average freezing point of seawater is about – 2 °C.

Dark red represents warmer temperatures and dark blue represents cooler temperatures.

4. Which latitude bands in each hemisphere contain the warmest surface waters? Which contain the coldest surface waters?
   a. Warmest —
   b. Coldest —

5. Given the underlying principle that heat flows from warmer to cooler areas, and knowing the direction that the gyres flow in each hemisphere, predict which boundary current—the eastern or western boundary current—will be warmer in each hemisphere. (Refer to Figure 2 and Map 1 on the previous page.)
   a. Northern Hemisphere *(circle one)*: western / eastern
   b. Southern Hemisphere *(circle one)*: western / eastern

To check your prediction, you will calculate the average temperature of the eastern boundary currents in the Northern and Southern Hemispheres. The western boundary currents have already been done for you and are listed in Table 1 on the following page.

- Click the Select By Attributes button.
- To select the eastern boundary currents in the Northern Hemisphere, query the **Water Temperature** layer for **("Type" = 'eastern boundary') and ("Hemisphere" = 'Northern')** as shown in steps 1-6. The query will actually read:

("CURRENT_TY" = 'eastern boundary') AND ("HEMISPHERE" = 'Northern')

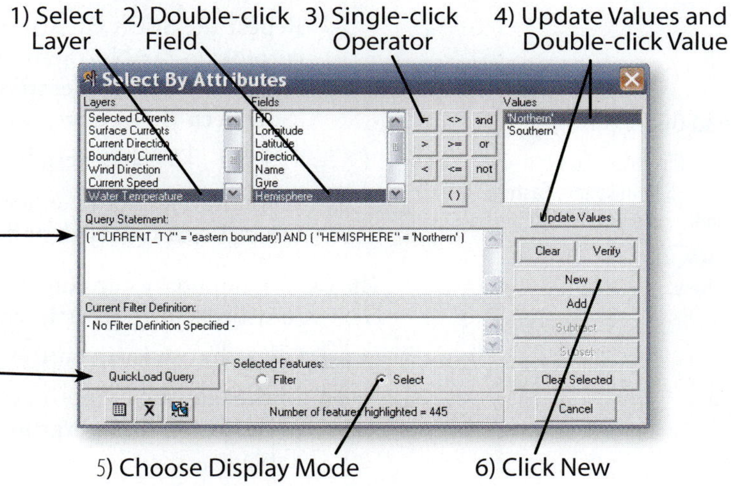

Read query statement here as you enter it.

QuickLoad Query

- Click the QuickLoad Query button and select the **N Hemisphere Eastern Boundary** query.
- Click **OK**.
- Click **New**.

- If you have difficulty entering the query statement correctly, refer to the **QuickLoad Query** described at left.

The eastern boundary currents in the Northern Hemisphere should be highlighted. Next you will calculate statistics for the selected data.

- Click the Statistics button in the Select By Attributes window.
- In the Statistics window, calculate statistics for **only selected features** of the **Water Temperature** layer, using the **Temp (C)** field.

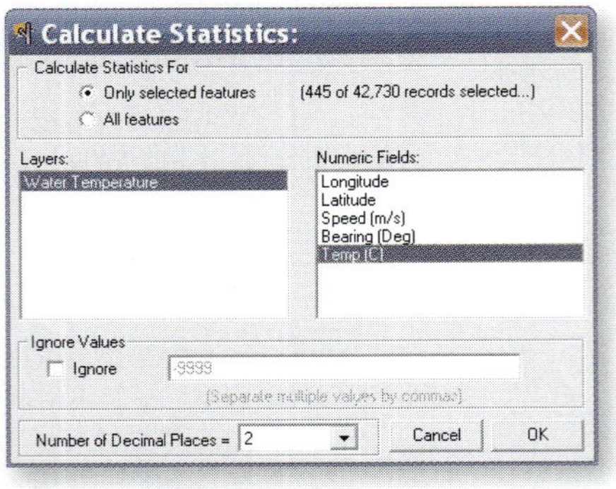

Click **OK**.

The average water temperature of the eastern boundary currents in the Northern Hemisphere is reported in the Statistics window as the **Mean**.

6. Record the average (**Mean**) temperature of the eastern boundary currents in Table 1. Round values to the nearest 0.1 °C.

Rounding

To learn more about rounding, see the section on *Rounding* in the Introduction to this guide.

Table 1 — Average temperature of boundary currents by hemisphere

Boundary currents	Average temperature (°C)	
	Northern Hemisphere	Southern Hemisphere
western	22.2	19.6
eastern		

Close the Statistics window.

Click **Clear** in the Select By Attributes window to erase the previous query.

Repeat the Select By Attributes operation to select the eastern boundary currents in the Northern Hemisphere. Query the **Water Temperature** layer for **("Type" = 'eastern boundary') and ("Hemisphere" = 'Southern')**. The query will actually read:

("CURRENT_TY" = 'eastern boundary') AND ("HEMISPHERE" = 'Southern')

If you have difficulty entering the query statement correctly, refer to the **QuickLoad Query** described at left.

QuickLoad Query

- Click the QuickLoad Query button and select the **S Hemisphere Eastern Boundary** query.
- Click **OK**.
- Click **New**.

The eastern boundary currents in the Southern Hemisphere should be highlighted. Next you will calculate statistics for the selected data.

Click the Statistics button [X̄] in the Select By Attributes window.

In the Statistics window, calculate statistics for **only selected features** of the **Water Temperature** layer using the **Temp (C)** field and complete Table 1.

7. How do these results compare with your predictions about the temperatures of boundary currents in question 5? Explain the patterns you see.

8. Use the temperature data in Table 1 and the direction of the currents you recorded on Map 1 to complete these statements about how boundary currents redistribute heat energy within oceans. Refer to Figure 2 (page 50) to review the four types of boundary currents.
   a. Western boundary currents transport *(circle one)* warm / cold water to *(circle one)* tropical / polar regions.
   b. Eastern boundary currents transport *(circle one)* warm / cold water to *(circle one)* tropical / polar regions.

- Close the Statistics window.
- Close the Select By Attributes window.
- Click the Clear Selected Features button.
- Turn off the **Water Temperature** layer.

Wind and current directions

Normally, currents are labeled according to the direction they are flowing *toward*, whereas winds are labeled according to the direction they are blowing *from*. To avoid confusion, winds and currents are both labeled according to the direction they are moving *toward* in this activity. Thus, a wind or current flowing from south to north is designated *N* or *north*.

Major ocean currents may also be named for their geographic location. For example, the Benguela (ben-GWAY-luh) Current is named after the port city of Benguela, on the coast of western Angola, Africa.

Speed patterns

Just as we can predict ocean temperatures based on where a current originates, the location of a current within a gyre also provides clues about the current's speed. Looking at the speed of the current may also help us answer the question "What drives the currents?"

- Turn on the **Current Speed** layer.
- Select the **Current Speed** layer.

This layer displays the average annual speed of water circulating in the world's oceans. Dark red represents higher speed (faster currents), whereas pink represents lower speed (slower currents). Examine the map and look for patterns in the current speed.

9. Which of the four types of gyre currents (eastern boundary, western boundary, eastern transverse, western transverse) appear to be moving the fastest in each hemisphere?

Next, you will examine the speed of currents in gyres by summarizing the speed data based on current type.

- Click the Summarize button **Σ**.
- In the Summarize window, select **Current Speed** as the **feature layer**.
- Select **Type** as the **field to summarize** in the drop-down menu.
- Double-click **Speed (m/s)** to display the statistics options and check **Average**, and click **OK.**

10. Use the summary table to completeTable 2 on the following page. Round values to the nearest 0.01 m/s.

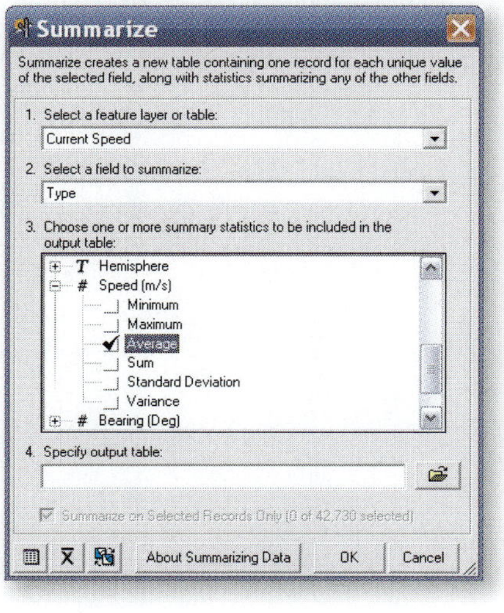

Table 2 — Average speeds of boundary current types

Boundary current type	Average speed *m/s*
Eastern boundary	
Northern transverse	
Southern transverse	
Western boundary	

Close the Summary Table window.

In the next section, you will explore why these currents flow at different speeds.

Where the wind blows

Global winds are general, consistent patterns of air movement driven by the sun's heat energy and Earth's rotation. Next, you will examine these circulation patterns and compare them to ocean surface-current patterns.

Turn off the **Current Speed** and **Ocean Labels** layers.

Turn on the **Wind Direction** layer.

11. Examine the predominant wind direction within each latitude band. Summarize your observations in Table 3. If there is not a clear overall movement of the wind in one direction, record the wind direction as **mixed**. (Important: Read the clarification about wind and current directions in the sidebar before you fill in the table.)

Wind and current directions

Normally, currents are labeled according to the direction they are flowing *toward*, whereas winds are labeled according to the direction they are blowing *from*. To avoid confusion, winds and currents are both labeled according to the direction they are moving *toward* in this activity. Thus, a wind or current flowing from south to north is designated *N* or *north*.

Major ocean currents may also be named for their geographic location. For example, the Benguela Current is named after the port city of Benguela, on the coast of western Angola, Africa.

Table 3 — Wind direction by latitude

Latitude band	Predominant wind direction
60° – 90° N	
30° – 60° N	
0° – 30° N	
0° – 30° S	
30° – 60° S	
60° – 90° S	

12. How does the wind direction change between 30° – 60° N latitudes as the winds approach the western edge of both North America and North Africa?

13. Does the same change in wind direction occur between 30° – 60° S as the winds approach the western edge of South America, southern Africa, and southern Australia? If not, how does the pattern differ?

14. Based on the currents you drew on Map 1 (page 50), how do the ocean current directions compare to the predominant wind — the same, opposite, or at some other angle to the wind direction? Explain your answer.

15. Using your knowledge of winds and the location of large land masses, explain why you think transverse currents flow faster than boundary currents.

Surface currents and winds

Surface currents appear to be related to prevailing winds, but do the winds and currents move in exactly the same direction? Next, you will examine four currents to compare the wind and the current direction.

- Turn on the **Selected Currents** layer.
- Select the **Selected Currents** layer.

The **Selected Currents** layer outlines large segments of four major surface currents: the California, Benguela, North Equatorial, and South Indian currents. You will be gathering and recording data about these currents in Table 4. Data for the Benguela and North Equatorial currents have been entered in the table for you.

16. For the California and South Indian current regions, visually estimate the predominant wind direction and record your answer in the **Direction - Wind** column of Table 4. (Note: It may help to turn the **Selected Currents** layer on and off.)

- Turn off the **Wind Direction** layer.
- Turn on the **Current Direction** layer.

The **Current Direction** layer shows the average direction of oceanic surface currents.

17. Visually estimate the predominant current direction for these two regions and record it in the **Direction - Current** column of Table 4. (Note: It may help to turn the **Selected Currents** layer on and off.)

Table 4 — Direction of global winds and surface currents

Surface current	Direction		Bearing		Wind-current offset direction	Hemisphere
	Wind	Current	Wind	Current		
Benguela	N	N	344	331	CCW	S
California						
North Equatorial	W	W	237	275	CW	N
South Indian						

Note: At this point, only the **Direction** section of Table 4 will be completely filled in for all four currents. You will complete the rest of the table later.

18. Compare the general direction of the currents to the direction of the winds in Table 4.

- Turn off the **Current Direction** layer.
- Turn on the **Wind Direction** layer.
- Select the **Selected Currents** layer.

You will return here to repeat this procedure

Next, you will compare your visual observation with each current's average direction (bearing) in degrees.

- Click the Select By Attributes button.
- To select the California Current, query the Selected Currents layer for (**"Name" = 'California'**). The query will actually read:

  ("CURRENT_NA" = 'California')

- If you have difficulty entering the query statement correctly, refer to the **QuickLoad Query** described at left.
- Click New.
- Close the Select By Attributes window.

QuickLoad Query

- Click the QuickLoad Query button and load the **California Current** query.
- Click **OK**.
- Click **New**.

The outlined region of the California Current should now be highlighted. Next, you will perform a Select By Location operation to determine the bearing of the wind over the California Current.

- Click the Select By Location button.
- In the Select By Location window, construct the query statement:

  I want to **select features from** the **Wind Direction** layer that **intersect** the features of the **Selected Currents** layer.

- Check the **Use selected features** box.
- Click **Apply**.
- Close the Select By Location window.
- Click the Statistics button.
- In the Statistics window, calculate statistics for **only selected features** of the **Wind Direction** layer, using the **Bearing (Deg)** field.
- Click **OK**.

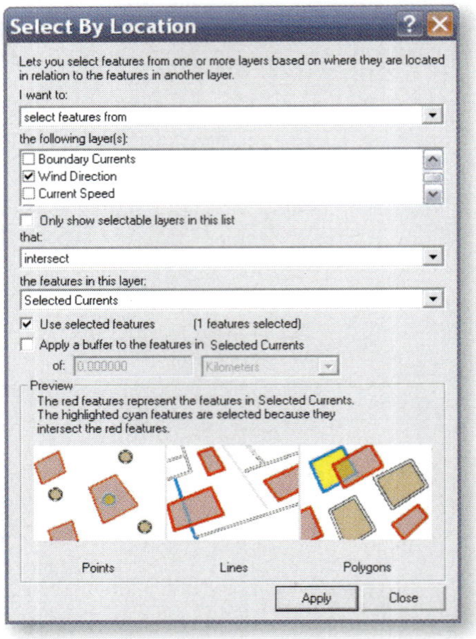

The average bearing of the wind above the California current is reported in the Statistics window as the **Mean**.

19. Round the average (**Mean**) wind bearing to the nearest degree and record it in the **Bearing - Wind** column of Table 4 on the previous page.

- Close the Statistics window.

Now you will determine the average direction (bearing) of the California Current in the outlined region.

- Click the Select By Location button.
- In the Select By Location window, construct the query statement:

  I want to **select features from** the **Current Direction** layer that **intersect** the features of the **Selected Currents** layer. (Note: Make sure that the **Current Direction** layer is the only box checked.)

- Check the **Use selected features** box.

- Click **Apply**.
- Close the Select By Location window.
- Click the Statistics button [X].
- In the Statistics window, calculate statistics for **only selected features of** the **Current Direction** layer, using the **Bearing (Deg)** field.
- Click **OK**.

The average bearing of the water within the California current is reported in the Statistics window as the **Mean**.

20. Record the average (**Mean**) current bearing in degrees in the **Bearing - Current** column of Table 4.

- Close the Statistics window.
- Click the Clear Selected Features button.
- Repeat the Select By Attributes, Select By Location, and Statistics operations for the **South Indian** Current and record the wind and current bearings in Table 4. (Hint: Return to the spot marked with a green arrow in the sidebar on the previous page to review the directions.)

QuickLoad Query

- Click the QuickLoad Query button and load the **South Indian Current** query.
- Choose the **Select** option.
- Click **New**.

21. On Graph 1, plot each surface current from Table 4. (The Benguela Current has been done for you.)
    a. Draw a solid line to indicate the average wind bearing.
    b. Draw a dashed line to indicate the average current bearing.
    c. Draw an arrow from the wind bearing to the current bearing. This arrow represents the *wind-current offset*, the difference between the direction the current is moving and the direction the wind is moving.
    d. Label the current.

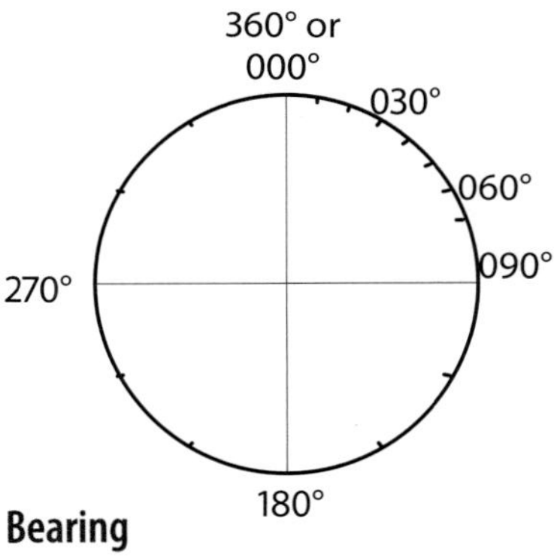

Bearing

Bearing is the location or direction of movement of something, calculated by using a map or compass. The units of bearing are in degrees. 000° and 360° are to the north, 090° is to the east, 180° is to the south, and 270° is to the west.

Graph 1 — Winds and currents of major surface currents

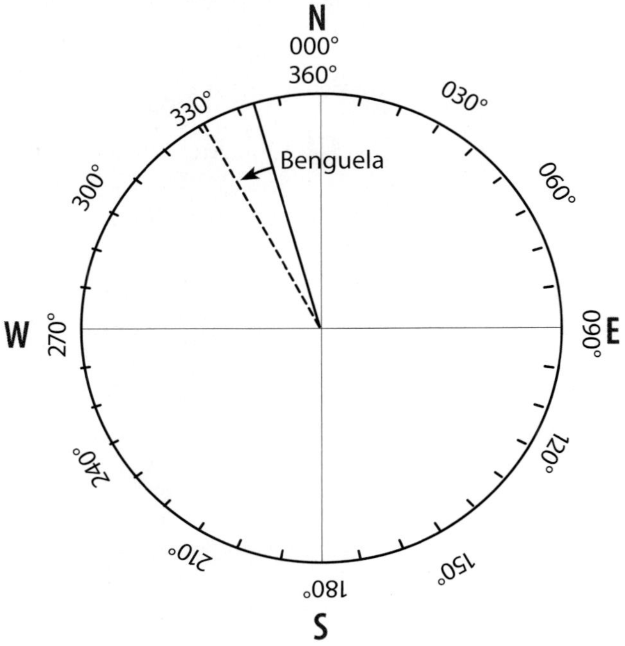

22. Examine the arrows you drew on Graph 1 from each wind to its related current and record whether the wind-current offset for each current is counterclockwise (CCW) or clockwise (CW) in the **Wind-current offset direction** column of Table 4.

23. Record the hemisphere where each current is located in Table 4. (Northern Hemisphere = N and Southern Hemisphere = S).

24. Using the data you recorded in Table 4, describe any patterns you see between the hemisphere and the wind-current offset direction.

    a. Northern Hemisphere —

    b. Southern Hemisphere —

25. What do you think causes the small difference in direction between a wind and its associated current? Do you think the wind is driving the current or is the current driving the wind? Explain.

Quit ArcMap and do not save changes.

Reading 2.3

Current basics

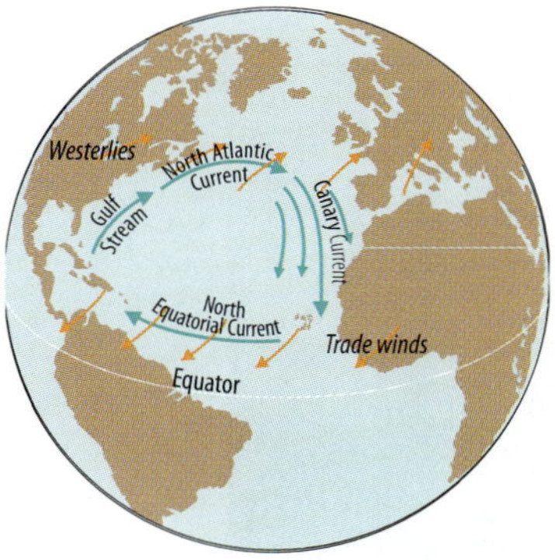

Figure 1. Global winds (orange) and their corresponding surface currents (blue) in the North Atlantic Ocean.

Ocean waters are continuously moving, circling the ocean basins in powerful currents hundreds of kilometers wide, and in swirls and eddies as small as a centimeter across. The primary forces driving the large-scale motions are the sun's energy and Earth's rotation. Energy from the sun warms Earth's surface and atmosphere, generating winds that initiate the horizontal movement of surface water (Figure 1 at left). Vertical movement between the surface and the ocean depths is tied to variations in temperature and salinity, which together alter the **density** of seawater and trigger sinking or rising of water masses. Together, the horizontal and vertical motions of water link the world's oceans in a complex system of surface and subsurface currents often referred to as the **Global Conveyor Belt** (Figure 2). This circulation system plays a vital role in transporting and distributing heat, nutrients, and dissolved gases that support life around the globe.

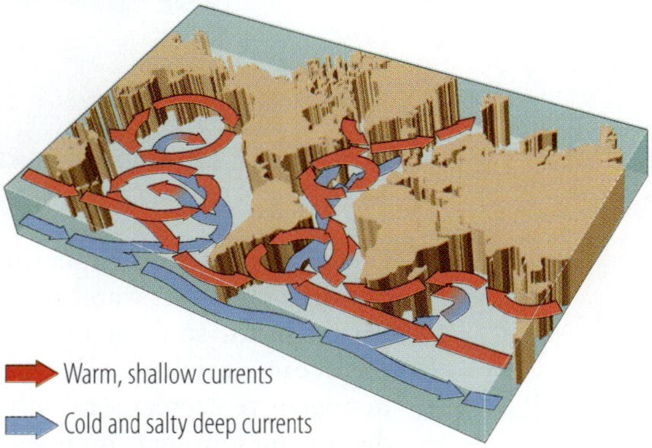

Figure 2. A highly simplified diagram of the Global Conveyor Belt.

Structure of the ocean waters

density — the mass per unit volume of a substance or object.

$$\text{density (kg/m}^3) = \frac{\text{mass (kg)}}{\text{volume (m}^3)}$$

The oceans contain numerous **water masses**, which can be recognized as different by their physical and chemical characteristics such as salinity, temperature, and density. The density of seawater depends on its temperature and salinity, as well as the amount of pressure exerted on it. Water expands as it warms, increasing its volume and decreasing its density. As water cools, its volume decreases and its density increases. Salinity, the amount of dissolved solids (like salts) in the water, alters density because the dissolved solids increase the mass of the water without increasing its volume. So, as salinity increases, the density of the water increases. Finally, when the pressure exerted on water increases, its density also increases.

Changing density

The density of water changes as its temperature or salinity (or both) change.

- If the temperature decreases and/or the salinity increases, the water becomes more dense.
- If the temperature increases and/or the salinity decreases, the water becomes less dense.

1. Rank the following types of ocean water from highest density (1) to lowest density (3).
   a. Warm, salty water _____
   b. Cold, salty water _____
   c. Warm, fresh water _____

photosynthesis — the process by which chlorophyll-containing plants convert sunlight and carbon dioxide to carbohydrates (food) and oxygen (O_2).

The characteristics of a water mass typically develop at the ocean surface due to interactions with the atmosphere. Evaporation can increase salinity as fresh water is removed from the ocean and the salts are left behind. Precipitation has the opposite effect, decreasing salinity levels as fresh water is added to the ocean. Processes like **photosynthesis** and the exchange of energy and matter between the ocean surface and the atmosphere can affect the amounts of oxygen and other dissolved gases in the water.

Thermocline

The thermocline is a layer of the ocean in which the temperature decreases rapidly with depth. Above the thermocline, the temperature is fairly uniform due to the mixing processes of currents and wave action. In the deep ocean below the thermocline, the temperature is cold and stable.

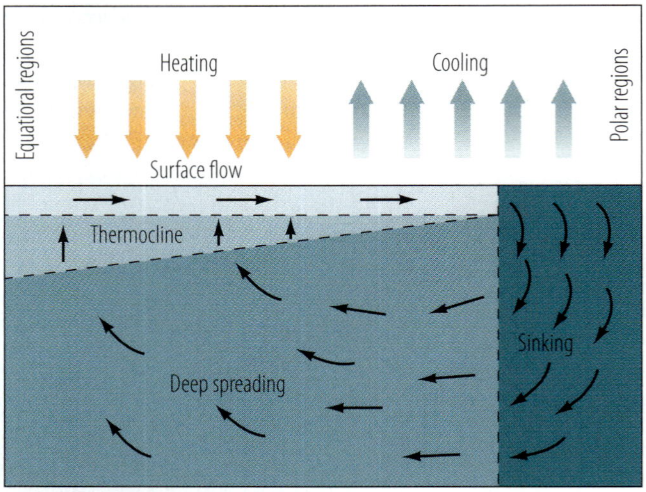

Figure 3. Schematic cross section of ocean from equator to pole.

In addition, water temperature (and thus density) changes rapidly as surface currents transport water masses from the equator to the poles and vice versa. Although the sun's energy is very efficient at warming the upper 100 meters of the ocean, very little solar energy penetrates to deeper waters. Therefore, water temperature decreases rapidly between 100 and 800 m depth. This region of decreasing temperature is called the **thermocline**, and marks the boundary between surface-water circulation and deep-water circulation (Figures 3 and 4).

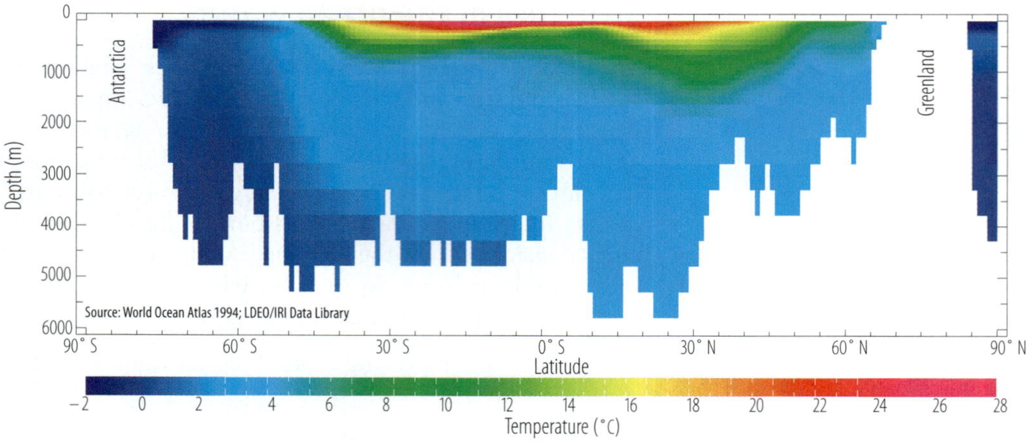

Figure 4. South-north temperature profile of the Atlantic Ocean at 32.5° W longitude. White represents the ocean floor and continents.

2. The water temperature at the base of the thermocline is around 5 °C. Using this information, sketch and label the approximate location of the base of the thermocline in Figure 4.

A similar zone, in which salinity changes rapidly with depth, is called the **halocline** (Figure 5). However, the halocline is not as well defined as the thermocline and in some places does not exist.

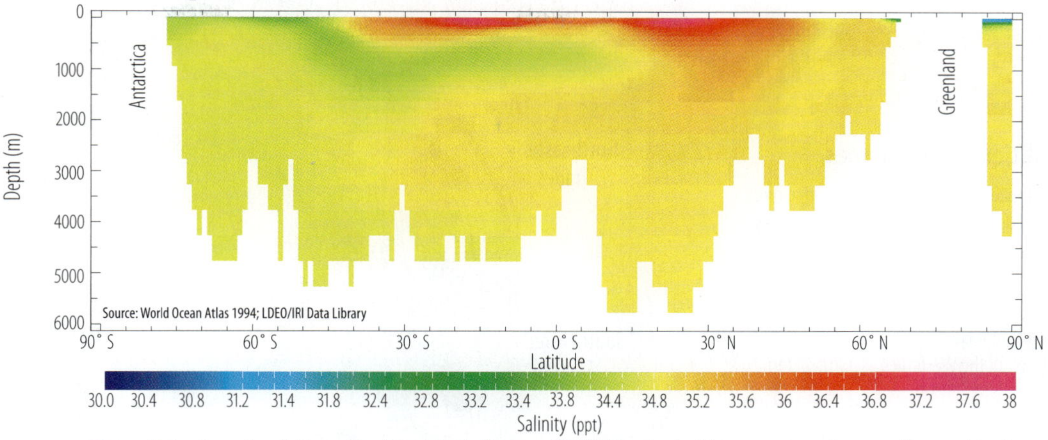

Figure 5. South-north salinity profile of the Atlantic Ocean at 32.5° W longitude. White represents the ocean floor and continents.

Once formed, water masses tend to retain their original characteristics because they mix very slowly with the surrounding water—except in places where the thermocline is very weak. Their distinctive characteristics make it possible to identify their place of origin and track their movements. In fact, it is by tracking differences in the physical properties of water masses that scientists have been able to begin mapping the Global Conveyor Belt.

Wind-driven currents

Winds are created by the uneven heating of Earth's surface by the sun, due primarily to Earth's nearly spherical shape (Figure 6). Surface temperature variations create temperature and pressure differences in the layer of air near the surface. To equalize these differences, air moves from regions of high pressure to regions of low pressure, creating wind.

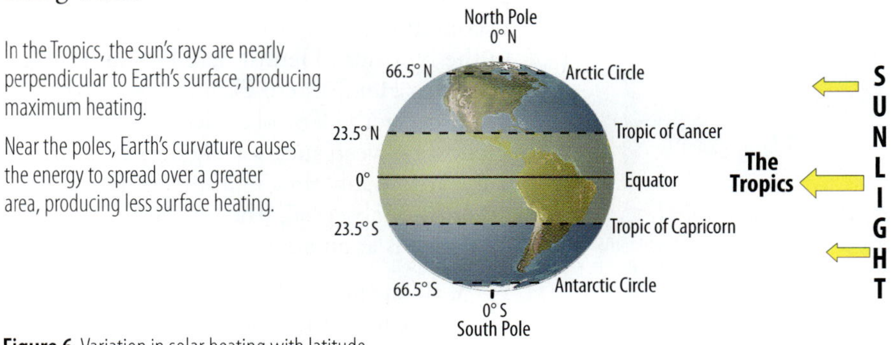

Figure 6. Variation in solar heating with latitude.

Spreading light

When the sun is directly overhead at the equator, the same amount of sunlight that falls on one square meter at the equator would be spread over two square meters in Anchorage, Alaska.

Low pressure belts form where warm air rises, near the equator and around 60° latitude (Figure 7 on the following page); high pressure belts are found where cool air sinks, near the poles and around 30° latitude. Air moving from high pressure toward low pressure creates six global wind belts encircling Earth. These belts shift slightly north and south with the seasons, but they are otherwise permanent features. Strong prevailing winds and solar warming produce ocean surface currents that extend to depths ranging from 45–400 m under typical conditions. This surface layer of currents is called the **Ekman layer**, or the **wind-blown layer**.

Traditional wind names

The global wind belts in Figure 6 are named, by tradition, according to the direction they are *blowing from*. In these materials we name both winds and ocean currents according to the direction they are *blowing toward*. For example, in the Northern Hemisphere, we would describe the direction of the Westerlies as *northeast* (or *NE*).

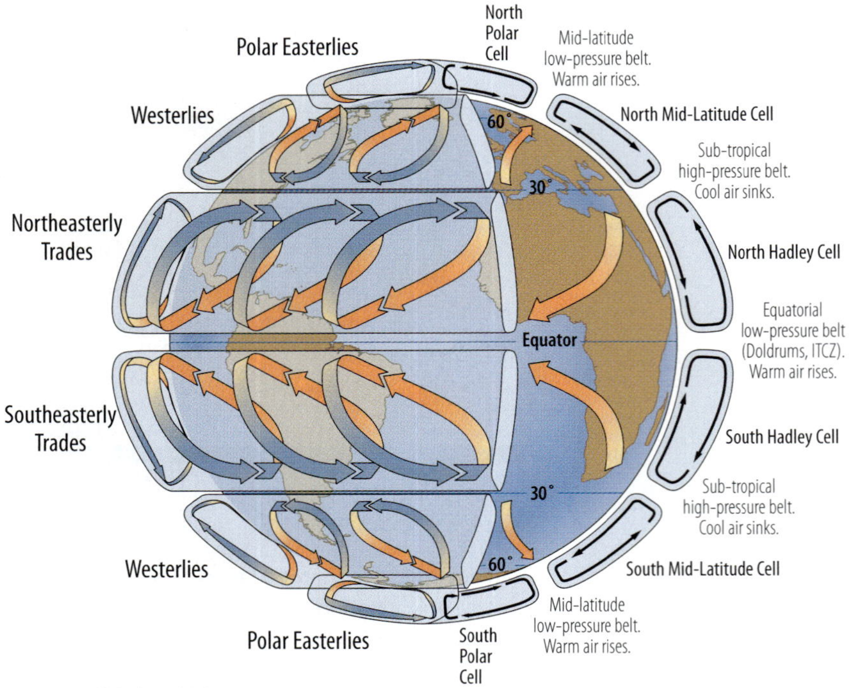

Figure 7. Global wind belts.

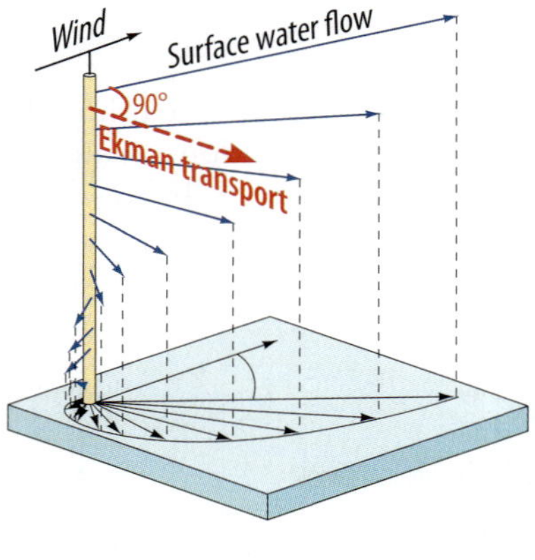

Figure 8. The Ekman spiral. The red arrow represents the net effect, called Ekman transport. Clockwise Northern Hemisphere deflection is shown here. (Southern Hemisphere deflection is counterclockwise.)

Note: The water does not spiral downward like a whirlpool.

The Coriolis effect and Ekman transport

Over short distances, winds and the ocean surface currents they generate follow straight paths, but over greater distances they curve due to Earth's rotation. This phenomenon is called the *Coriolis effect*. In the Northern Hemisphere, the Coriolis effect causes winds and ocean currents to veer to the right; in the Southern Hemisphere, the winds and ocean currents curve to the left.

As you learned in Investigation 2.2, Ekman transport is an offset between a current direction and its associated wind. It is useful to think of the Ekman layer as containing many thinner layers of water flowing over one another (Figure 8). In the Northern Hemisphere Ekman transport is deflected to the right and in the Southern Hemisphere Ekman transport is deflected to the left. This phenomenon is caused by the Coriolis effect and the slowing and deflection of water due to friction between successively deeper layers of water. It is theoretically possible for water to actually flow in a direction opposite to the surface current, but this has never been observed. The overall motion of the Ekman layer, referred to as *Ekman transport*, is at an angle of about 90° to the wind direction.

3. If the arrows below represent the prevailing winds somewhere over the ocean in the Northern and Southern Hemispheres, draw another arrow to indicate which direction Ekman transport would cause water to flow.

   a. Northern

   b. Southern

Wind-driven upwelling and downwelling

In nearshore environments, it is common to have winds blowing parallel to shore over the ocean (Figure 9). Ekman transport moves surface water offshore and pulls deep, cold, nutrient-rich water to the surface. This process, known as **wind-driven upwelling**, is restricted mainly to the west coast of continents, and is responsible for the high productivity of nearshore waters.

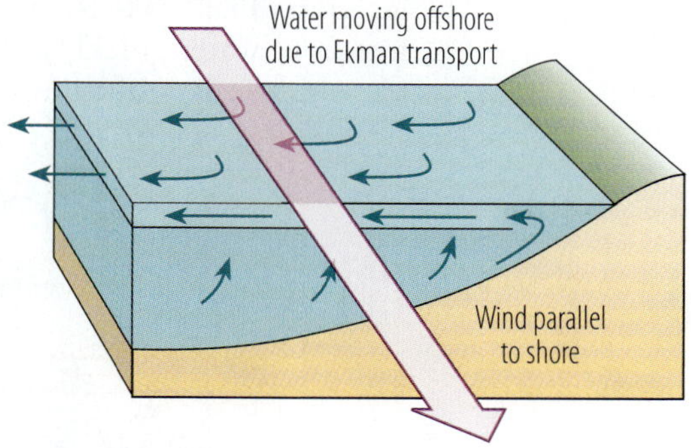

Figure 9. Factors that produce coastal upwelling.

Upwelling occurs in the open ocean near the equator in a similar manner (Figure 10). On both sides of the equator, surface currents moving westward are deflected slightly poleward and are replaced by nutrient-rich, cold water from great depths.

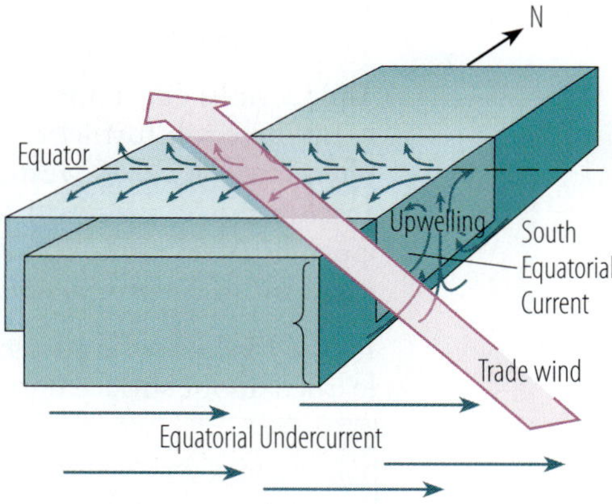

Figure 10. Factors that produce equatorial upwelling.

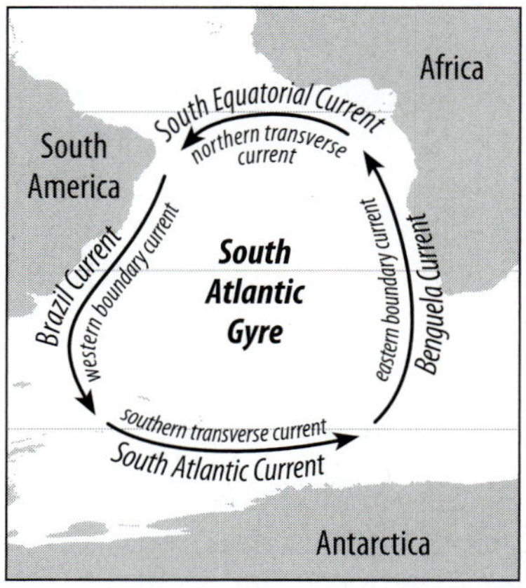

Figure 11. The four boundary currents that form a gyre.

The mechanical action of wind on the currents promotes mixing of the Ekman layer, which tends to deepen the thermocline and promote the upwelling of nutrients. The thermocline, which separates less dense, warm surface water from the more dense, cold water below, is most pronounced at low latitudes and prevents nutrient-rich deep waters from rising to the surface. In contrast, upwelling occurs more readily in high-latitude regions near the poles. These regions receive little sunlight and are not warmed by solar energy. Without a distinct thermocline, upwelling easily brings nutrients toward the surface and promotes mixing.

Surface currents

Gyres play a major role in redistributing the sun's heat energy around the globe. Each gyre consists of four interconnected, yet distinct currents (Figure 11). A pair of boundary currents flows north or south, parallel to the bordering landmasses.

Western boundary currents carry warm equatorial water poleward, while eastern boundary currents carry cooler temperate and polar water toward the equator. These currents interact with the air near the surface to moderate the climate of coastal regions. Within a gyre, boundary currents are connected by transverse currents. Transverse currents move east or west across the gyre's northern and southern edges.

The speed of a current within a gyre is related to the prevailing winds and the location of landmasses. Western boundary currents are narrow but move huge masses of water quickly as the westward-blowing trade winds push water against the eastern edges of continental landmasses (Figure 12).

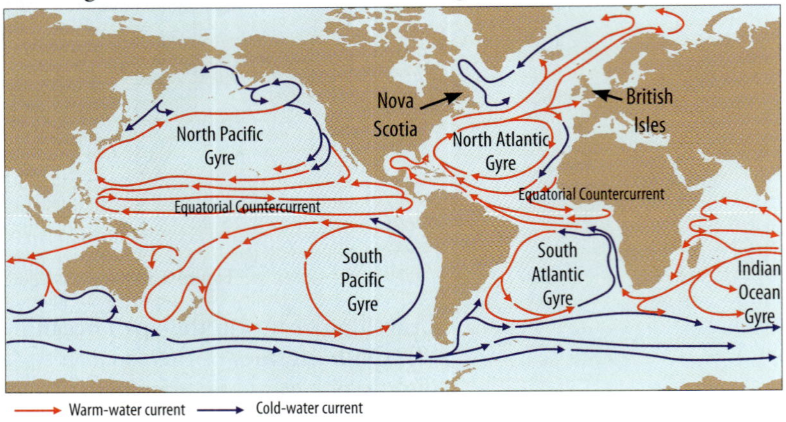

Figure 12. Major ocean surface currents and gyres.

The Coriolis effect and resulting Ekman transport occurring at 90° from the wind direction further enhance the speed of western boundary currents, a phenomenon called **western intensification**. Although most of the water at the equator moves westward then poleward, the low-intensity winds and lack of Coriolis effect at the equator allow for some of the water at the surface to flow eastward in equatorial countercurrents.

Figure 13 shows the currents of the North Pacific Gyre. Use what you have learned about surface currents to answer the following questions.

Kuroshio = kuhr-oh-SHEE-oh

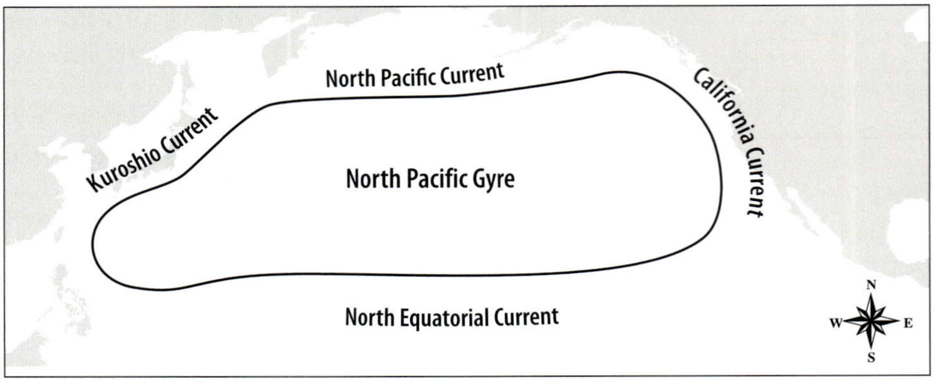

Figure 13. The North Pacific Gyre.

4. Draw arrows on the map to show the direction of the four currents labeled.

5. Complete Table 1 on the following page with information about the surface currents of the North Pacific Gyre. For **Heat exchange type**, indicate whether the current is gaining heat (warming) or losing heat (cooling) as it flows.

Table 1 — Boundary currents of the North Pacific Gyre

Type of current	Name	Heat exchange type *cooling / warming*
Eastern Boundary		
Western Boundary		
Northern Transverse		
Southern Transverse		

Density-driven currents

In addition to wind-driven horizontal surface currents, ocean circulation has a vertical component that is driven by differences in water density. When surface water cools or becomes more saline due to evaporation or other processes, its density increases and it sinks either to the bottom of the ocean or to a depth where its density equals that of the surrounding water. This density-driven circulation pattern is referred to as **thermohaline circulation**, and the currents it produces are called **density currents**. The cold water eventually returns to the surface to be reheated and returned to the poles by surface currents, or to mix with other water masses and return to the depths. Thermohaline currents move very slowly—about 1 centimeter per second—10 to 20 times slower than surface currents.

saline—salty [Latin *sal* = salt].

thermohaline—combined effects of temperature [Greek *thermo* = heat] and salinity [Greek *hal* = salt].

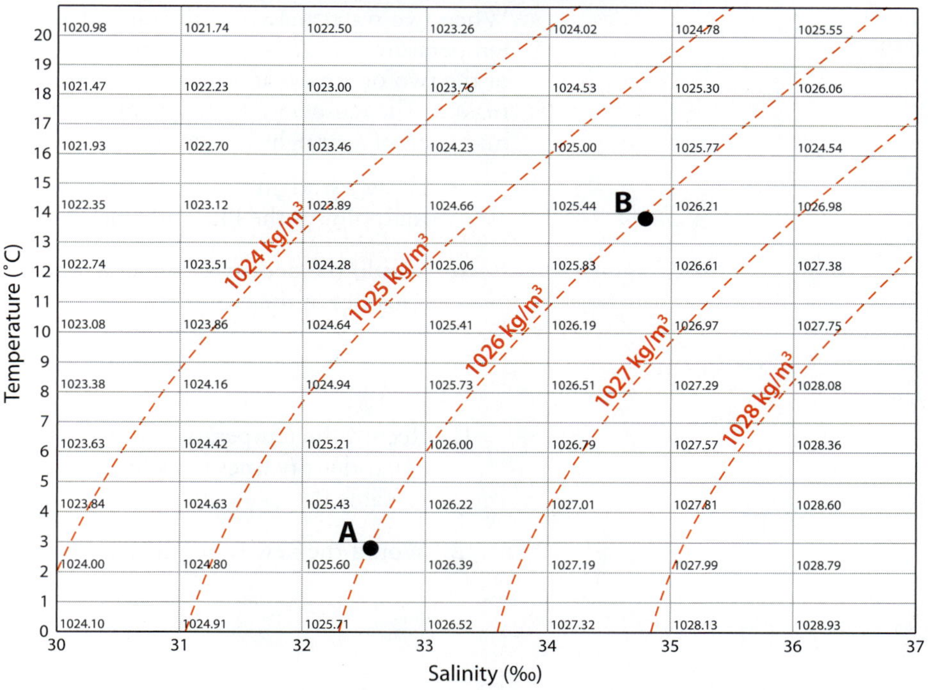

Figure 14. Ocean water Temperature-Salinity-Density chart showing the relationship between temperature, salinity, and density of ocean water. The dashed lines are lines of constant density.

Salinity

The average salinity of seawater is 34.7 ppt or **parts per thousand** (also symbolized ‰). That means that a liter of ocean water (a little more than a quart) contains 34.7 grams (~2.5 tablespoons) of various salts.

To learn more about the composition of seawater, click the Media Viewer button [button] and choose **Seawater**.

6. Examine Figure 14. What happens to the density of the water as the temperature decreases? (Follow one of the vertical lines of constant salinity downward, and note what happens to the density values.)

7. Use Figure 14 (on the previous page) to determine what happens to the density of ocean water as the salinity increases? (Follow one of the horizontal lines of constant temperature from left to right, and note what happens to the density values.)

Deep currents are generated by relatively small density variations. In fact, the density of seawater must be determined to several decimal places to detect significant differences. The points labeled A and B on Figure 14 represent the salinity and temperature values for two water masses.

8. Use Figure 14 to determine the temperature, salinity, and density of water masses A and B and record them in Table 2.

Table 2 — Mixing of water masses A and B

Point	Temperature *°C*	Salinity *ppt*	Density *kg/m³*
A			
B			
C			

When two water masses of the same density meet, they tend to mix. The temperature and salinity of the new water mass lie somewhere between those of the two original water masses. Imagine mixing equal parts of the two water masses. The temperature and salinity of the new water mass would lie at the midpoint of a straight line connecting point A to point B.

9. Draw a straight line connecting points A and B on Figure 14. Plot the midpoint of the line and label it C.

10. Would the density of the new water mass C be higher or lower than the densities of the two original water masses, A and B?

11. Record the temperature and salinity of point C in Table 2. Use the curved equal-density lines to estimate the density of water mass C and record it in Table 2.

12. Would the new water mass remain at the surface or sink? Explain.

Stability and instability of water masses

When the density of a water column increases with depth, the water column is **stable** and mixing does not occur. Conversely, when the density of a water column decreases with depth, it is **unstable**. As the dense water sinks, it produces turbulence and mixes with the layers beneath it. Instability is caused by an increase in the density of surface water due to a decrease in temperature, an increase in salinity, or both.

High evaporation rates can increase the salinity of the surface water; and low air temperatures can cool the surface water, causing it to become unstable and sink. When sea ice forms near the poles, most of the salt remains in the liquid water, increasing its density and producing instability.

There is also a seasonal aspect to ocean stability. During spring and summer, stability increases as the ocean surface warms. In fall and winter, stability decreases as the ocean surface cools.

Areas of instability can produce complex patterns of stratification and thermohaline and surface circulation in the ocean.

As sea ice forms along the coast of Antarctica, surface water cools and becomes more salty. This process is called brine rejection. This salty water sinks and flows northward along the ocean floor, forming the Antarctic Bottom Water mass (AABW). As winds blow the Antarctic Surface Water (AASW) eastward, the Coriolis effect deflects it toward the north. This causes upwelling of warmer, salty water, the Northern Atlantic Deep Water (NADW). This water mass mixes with the AASW to form the Antarctic Intermediate Water mass (AAIW). Because the AAIW is denser than the surface water (the Subantarctic Water mass or SAAW), it sinks below the SAAW at the Antarctic convergence.

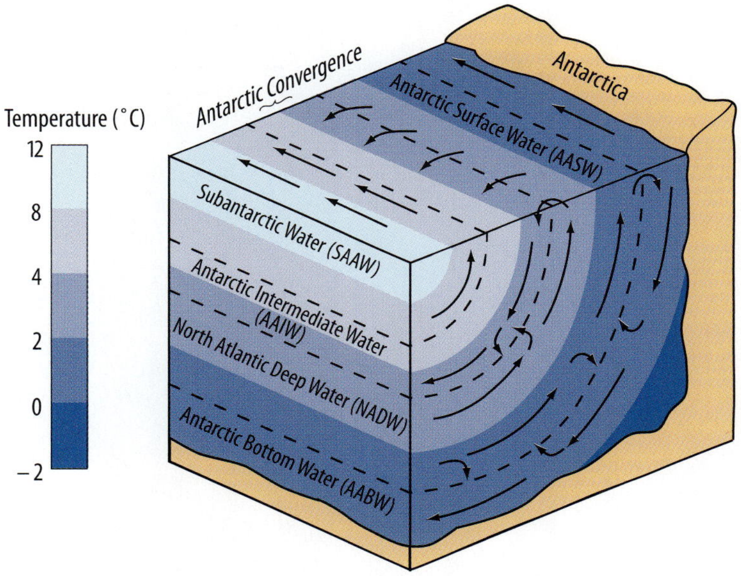

Figure 15. Thermohaline and surface currents off the coast of Antarctica. Colors represent water temperature, and dashed lines represent the boundaries between water masses.

13. Is the water column shown in Figure 15 stable or unstable? Explain.

Investigation 2.4

Deep-water currents

Deep-water currents, also called *density currents* or *thermohaline currents*, play a crucial role in maintaining the delicate balance of energy and nutrients in the marine environment. In this activity, you will examine two key factors that control water density—global temperatures and salinity patterns—to understand how they vary and how they affect where density currents form.

Water temperature

The first step in investigating the formation of density currents is to examine the average temperature of the ocean's surface. Temperature alters the density of seawater because water contracts when it cools. Thus, cooler water takes up less space or volume than warmer water. As a result, the density of water increases as water temperature decreases.

Launch ArcMap, and locate and open the **etoe_unit_2.mxd** file.

Refer to the tear-out Quick Reference Sheet located in the Introduction to this module for GIS definitions and instructions on how to perform tasks.

In the Table of Contents, right-click the **Deep-water Currents** data frame and choose Activate.

Expand the **Deep-water Currents** data frame.

Solar Flux movie

Another term for solar radiation is *solar flux*. To see an animation of how solar flux varies throughout the year, click the Media Viewer button and choose the **Solar Radiation** movie from the media list.

Notice that the poles receive more solar radiation during their respective summers — up to 24 hours a day — than at other times of year. However, the sunlight is spread over a larger area at the poles, reducing its intensity.

The **Avg Solar Radiation** layer shows the average amount of solar radiation per year, in watts per square meter, that would strike the surface if there were no clouds in the atmosphere to reflect the sunlight. Dark shades represent low amounts of radiation, and lighter shades indicate higher amounts.

1. Describe how the average solar radiation varies with
   a. latitude (from pole to pole).

   b. longitude (from east to west).

2. If there were no surface or deep-water currents circulating in the ocean, what effect might solar radiation have on the average temperature of the ocean waters?

Turn off the **Avg Solar Radiation** layer.

Turn on the **Avg Sea-Surface Temperature** layer.

This layer shows the average sea-surface temperature. Red represents warmer temperatures and blue represents cooler temperatures.

3. How well does the pattern of sea-surface temperature compare to the solar radiation?

4. Identify areas where surface currents may carry warm water into colder regions or cold water into warmer regions. On Map 1, place a C where sea-surface temperatures are cooler than expected and a W where they are warmer than expected. (Look for four or five of each.)

Map 1 — Temperature / solar radiation anomalies and salinity extremes

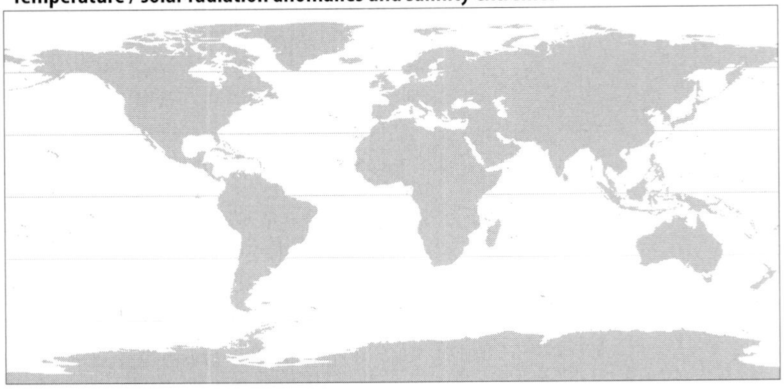

Sea-surface temperature is determined by many factors, such as solar radiation and the transfer of warm water from the equator to the poles and cold water from the poles to the equator. Deep-water convection also contributes to sea-surface temperature.

Turn on the **Upwelling** layer.

The **Upwelling** layer shows where deep-ocean currents rise to the surface, often becoming less dense than the surrounding waters.

5. On Map 1, mark places where upwelling occurs with the label **U**.

Compare the locations of the upwelling sites to their corresponding surface-temperature anomalies (if any).

6. Are upwelling sites more likely to be associated with cold or with warm surface-temperature anomalies? Explain.

Temperature anomalies

An *anomaly* is anything that is unusual, irregular, or abnormal. Sea-surface temperature anomalies occur where the surface temperature is warmer or cooler than expected for a given latitude.

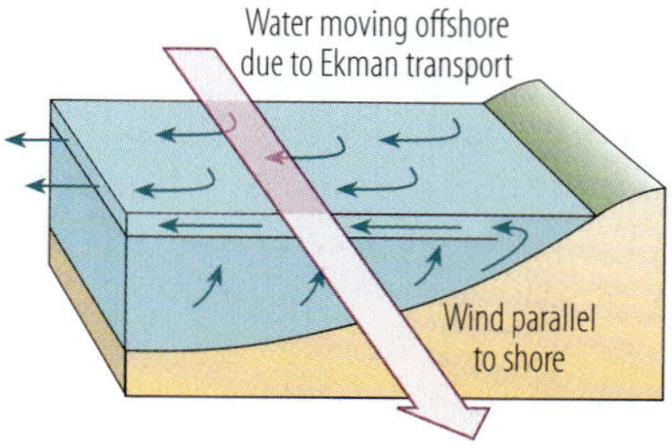

Figure 1. Conditions required for coastal upwelling.

Wind and current directions

Normally, currents are labeled according to the direction they are flowing *toward*, whereas winds are labeled according to the direction they are blowing *from*. To avoid confusion, winds and currents are both labeled according to the direction they are moving *toward* in this activity. Thus, a wind or current flowing from south to north is designated ***N*** or ***north***.

Major ocean currents may also be named for their geographic location. For example, the Benguela Current is named after the port city of Benguela, on the coast of western Angola, Africa.

Upwelling can occur as winds that blow parallel to the shoreline cause the surface waters to move away from the coastline, due to Ekman transport (Figure 1). Next you will examine whether winds play a role in developing these particular upwelling sites.

- Turn off the **Avg Sea-Surface Temperature** layer.
- Turn on the **Wind Direction** layer.

The **Wind Direction** layer displays the compass direction of the surface winds over both land and ocean.

- Examine the **Wind Direction** layer to determine the direction of the winds over the boundary currents. (Important: Read the clarification about wind and current directions in the sidebar.)

7. On Map 1 on the previous page, mark the upwelling sites that have winds moving parallel to shore with **PW** (parallel winds).

8. Describe any patterns you observe between the upwelling sites and the coastal winds.

Upwelling sites are very important for bringing nutrient-rich water from the deep ocean to the surface where marine life can flourish.

- Turn off the **Upwelling** layer.
- Turn off the **Wind Direction** layer.
- Turn on the **Density-Driven Downwelling** layer.
- Turn on the **Avg Sea-Surface Temperature** layer.

The **Density-Driven Downwelling** layer shows where surface water sinks as it becomes more dense than the surrounding waters. It will sink to a depth at which all the water below it is more dense and the water above is less dense. Next, you will compare the locations of density-driven downwelling sites to their corresponding surface-water temperatures.

- Click the Identify tool.
- In the Identify Results window, select the **Avg Sea-Surface Temperature** layer from the list of layers.
- Next, click within each of regions displayed by the **Density-Driven Downwelling** layer. The water temperature is listed in the **Temp (C)** field under **Value**.

9. In what range of surface-water temperatures are you most likely to find density-driven downwelling sites?

- Close the Identify Results window.

Global ocean circulation requires that if water is rising in one area, it must be sinking in another area. Next you will continue to explore the factors that influence the location of deep-water convection sites.

- Turn off the **Density-Driven Downwelling** layer.
- Turn off the **Avg Sea-Surface Temperature** layer.

Salinity

The average salinity of seawater is 34.7 ppt or *parts per thousand* (also symbolized ‰). That means that a liter of ocean water (a little more than a quart) contains 34.7 grams (~ 2.5 tablespoons) of various salts.

To learn more about the composition of seawater, click the Media Viewer button and choose **Seawater**.

Global salinity patterns

Salinity, in addition to temperature, strongly influences water density. The salts dissolved in seawater increase the density by adding to the water's mass without changing its volume. Next, you will explore global patterns of ocean salinity and the factors that influence this important property of seawater.

- Turn on the **Ocean Salinity** layer group.
- Select the **Avg Annual Salinity** layer.

This layer group displays the average annual salinity values for the world's oceans as well as some labels that you will examine later. Higher salinity is shown in shades of green, whereas lower salinity is shown in shades of blue.

- Turn on the **Major Rivers** layer.
- Use the Zoom In tool to examine the average annual salinity data. Look for patterns and anomalies in the data.
- Examine the regions with the highest and lowest salinity. Be sure to look around the edges of the continents as well as in the smaller seas.

10. Describe how the average salinity varies with

    a. latitude (from pole to pole).

    b. longitude (from east to west).

11. What factors might account for the salinity levels in the areas of highest and lowest salinity? (Hint: Look at the different layers in the Table of Contents for possible factors.)

- Click the Full Extent button to view the entire map.
- Turn on the **Net Annual Evaporation** layer.

This layer shows the overall loss or gain of freshwater from Earth's surface in centimeters per year. Evaporation, in which liquid water is converted to a gas, moves freshwater from the ocean to the atmosphere. When ocean water evaporates, the salt dissolved in the water is left behind, increasing the ocean's salinity. Precipitation removes freshwater from the atmosphere and delivers it back to the ocean, reducing the ocean's salinity. Net evaporation is calculated by subtracting the total annual precipitation from the total annual evaporation.

Net Evaporation = Total Evaporation (E) – Total Precipitation (P)

12. Predict how you think net evaporation will affect salinity: Areas with positive net evaporation will have (circle one) higher / lower salinity than areas with negative net evaporation.

13. Where do you find areas with the

    a. *highest* net evaporation? Why might they be found there?

    b. *lowest* net evaporation? Why might they be found there?

Figure 2 shows the relationship between net evaporation and ocean salinity between 60° N and 60° S. The dip near the middle corresponds to the Intertropical Convergence Zone (ITCZ), a band of moist, unstable air that circles the globe around 7° N latitude. The high net evaporation near 30° N and 30° S is due to global bands of high pressure. The cool, sinking air in these bands produces clear skies and dry conditions. Most of the world's deserts are found at these latitudes.

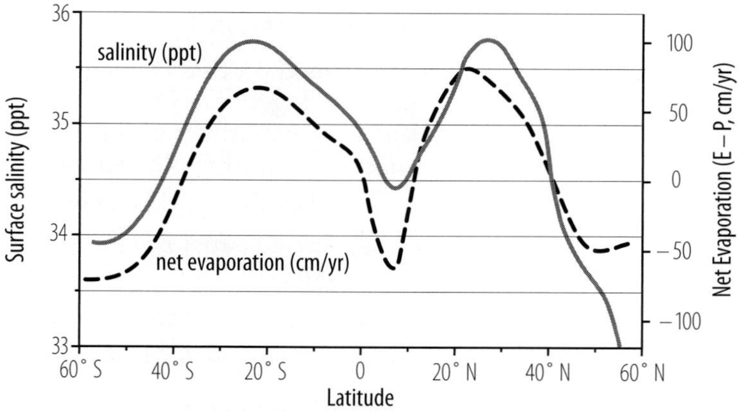

Figure 2. Net evaporation and surface salinity by latitude.

14. How well do the data in Figure 2 compare to the prediction you made in question 12?

- Turn off the **Net Annual Evaporation** layer.
- Turn off the **Major Rivers** layer.
- Turn on the **Density-Driven Downwelling** layer.
- Click the Identify tool.
- In the Identify Results window, select the **Avg Annual Salinity** layer from the list of layers.
- Next, click within each of regions displayed by the **Density-Driven Downwelling** layer.

15. What range of salinity correlates with the density-driven downwelling sites?

- Close the Identify Results window.

Seventy-five percent of the ocean has a salinity between 33 and 35 ppt. Thus, the salinity at density-driven downwelling sites is not particularly unusual. However, the temperature of seawater in these downwelling sites is definitely colder than the average sea-surface temperature. How does one water mass get colder and saltier than the waters around or below it? You will explore this question by following a water molecule from the saline mid-latitude waters to the polar region of the Atlantic Ocean.

Forming deep-water masses

Evaporation creates ocean-water masses with high salinity, but they are too warm to sink as density currents. In this section, you will investigate how cold, dense waters are generated.

The **Density-Driven Downwelling** layer shows several regions with significant downwelling in the North Atlantic and Southern Oceans. By measuring the surface characteristics in and around these regions, you will better understand the conditions under which density currents form.

- Click the QuickLoad button.
- Select **Spatial Bookmarks**, choose **North Atlantic** from the list, and click **OK.**
- Turn on the **Salinity Measurements** layer.
- Turn on the **Surface Currents** layer.

The **Surface Currents** layer shows the direction and temperature characteristics of surface currents in the North Atlantic Ocean.

- Click the Identify tool.
- In the Identify Results window, select the **Avg Annual Salinity** layer from the list of layers.
- Next, click points labeled B – E and record the temperature and salinity data in Table 1 on the following page. (Point A has been done for you.)

Table 1 — Surface water characteristics in the North Atlantic

Point	Temperature °C	Salinity *ppt*	Density *kg/m³*
A	17.5	35.6	1026.0
B			
C			
D			
E			

16. On Figure 3, plot and label the temperature and salinity values you recorded for points B through E. Use the curved dotted lines on the chart to determine the density of the surface water at each point, and record the water density for each point in Table 1.

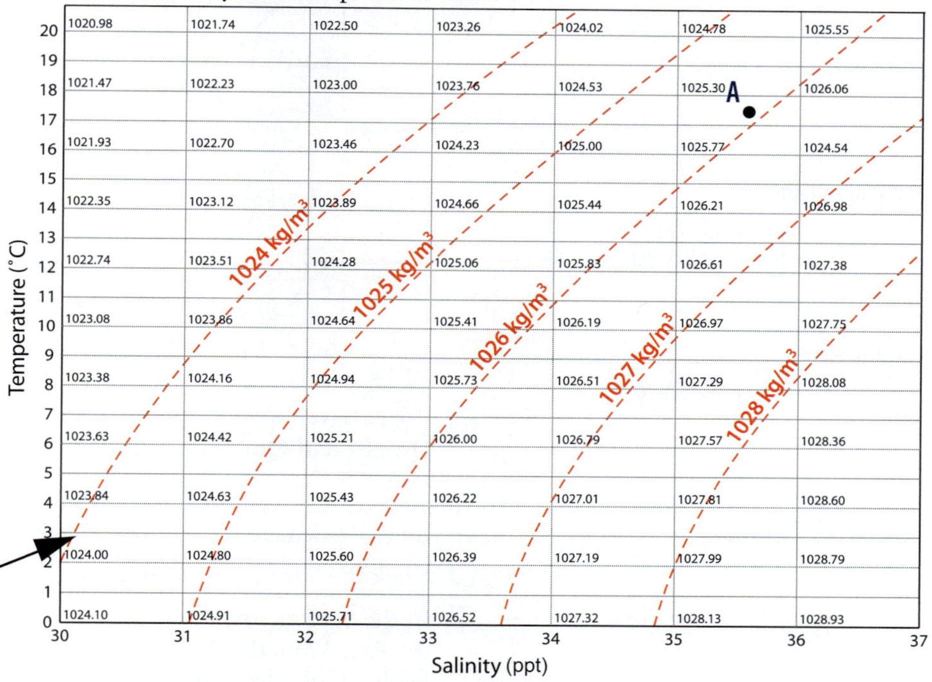

Each of the arcing dashed lines follows a constant density. Every point on this red dashed line has a density of 1024 kg/m³.

Figure 3. Temperature-Salinity-Density chart for seawater.

17. How does the density change from point A to point E?

How dense is freshwater?

Under normal sea-level temperature and pressure, pure fresh water has a density of 1000 kg/m³.

- Close the Identify Results window.
- Turn off the **Surface Currents** layer.

To cause density-driven downwelling, the surface water must be denser than the water beneath it, causing the layers to overturn. This occurs in the North Atlantic Ocean, where fresh meltwater from sea ice and the Greenland ice sheet form low-density surface water masses. As warm, high-salinity surface currents flow northward into this region, they cool rapidly and become denser. When these two water masses meet in the North Atlantic Ocean, the water column becomes unstable and the higher-density, high-salinity current sinks.

The densest waters in the ocean range from 1027 to 1029 kg/m^3. These correspond to a temperature of 6 °C or less and a salinity of 34 ppt or higher. Next, you will perform a series of operations to identify where those waters lie and compare their density to the waters surrounding them.

- Click the Select By Attributes button.
- To select the densest surface waters in the Northern Hemisphere, query the **Avg Annual Salinity** layer for **("Hemisphere" = 'N') and ("Temp (C)" <= 6) and ("Salinity (ppt)" >= 34)** as shown in steps 1-6 below. The query will actually read:

  ("HS" = 'N') AND ("TEMP_C" <= 6) AND ("SALINITY" >= 34)

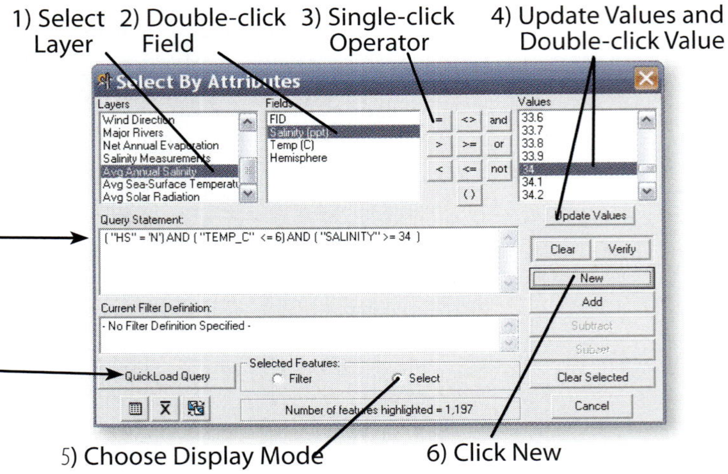

Read query statement here as you enter it.

QuickLoad Query

- Click the **QuickLoad Query** button and load the **NH Dense Surface Waters** query.
- Click **OK**.
- Click **New**.

- If you have difficulty entering the query statement correctly, refer to the **QuickLoad Query** described at left.

The densest surface waters of the Northern Hemisphere should be highlighted. Next, you will calculate statistics for the selected data.

- Click the Statistics button in the Select By Attributes window.
- In the Statistics window, calculate statistics for **only selected features** of the **Avg Annual Salinity** layer, using the **Temp (C)** and **Salinity (ppt)** fields. (Hold down the shift key to select more than one field.)

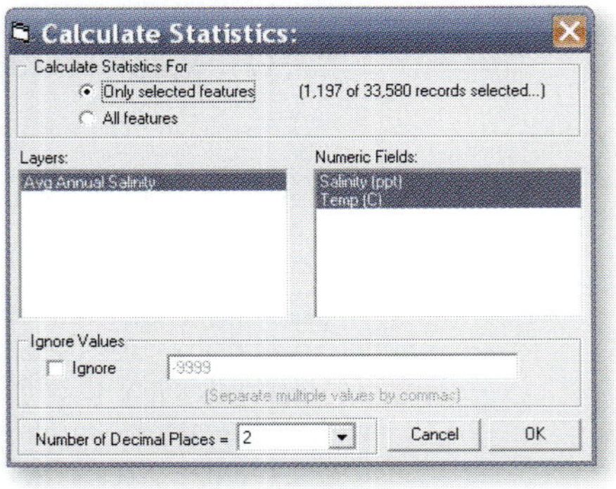

- Click **OK**.

The average (**Mean**) salinity and temperature inside the deep-water convection areas is reported in the Statistics window as the **Mean**.

18. Record the **Mean** salinity and temperature inside the deep-water convection areas in Table 2. Round values to the nearest 0.1 units. Use these values to determine the average density from Figure 2 and record the density in Table 2.

Table 2 — Surface-water characteristics of downwelling areas

Area	Temperature *℃*	Salinity *ppt*	Density *kg/m³*
Deep-water convection areas			
Adjacent waters			

- Close the Statistics window.
- Close the Select By Attributes window.
- Click the Clear Selected Features button.
- Turn on the **Adjacent Waters** layer.
- Click the Select By Location button.
- In the Select By Location window, construct the query statement:

  I want to **select features from** the **Avg Annual Salinity** layer that **intersect** the features in the **Adjacent Waters** layer.

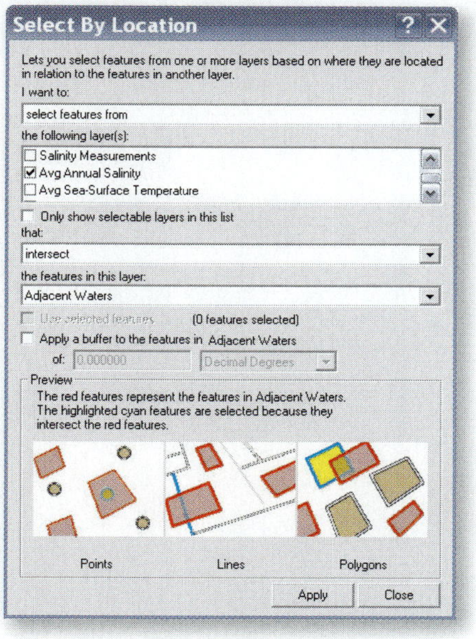

- Click **Apply**.
- Close the Select By Location window.
- Click the Statistics button.
- In the Statistics window, calculate statistics for **only selected features** of the **Avg Annual Salinity** layer, using the **Temp (C)** and **Salinity (ppt)** fields. (Hold down the shift key to select more than one field.)
- Click **OK**.

The average salinity and temperature for the adjacent waters is reported in the Statistics window as the **Mean**.

19. Record the average (**Mean**) salinity and temperature for the adjacent waters in Table 2. Use these values to determine the average density from Figure 2 and record the density in Table 2.

Close the Statistics window.

20. Compare the density of water in the deep-water convection areas with that of the adjacent waters. Explain how this relates to the formation of density currents.

Click the Full Extent button to view the entire map.

The deep-water convection area you have been studying is known as the North Atlantic Deep Water (NADW) current. Another major deep-water convection area is located in the Weddell (wuh-DELL) Sea, off the coast of Antarctica, southeast of South America. This area is the source of the Antarctic Bottom Water (AABW) density current. These deep-water density currents play a major role in the distribution of nutrients throughout the world's oceans.

Quit ArcMap and do not save changes.

Wrap-up 2.5

Stopping the flow

The North Atlantic Deep Water (NADW) and Antarctic Bottom Water (AABW) currents are two important parts of the Global Conveyor Belt. The volume of the NADW, which forms as the warm, salty Gulf Stream current moves north, cools, and sinks, is believed to have a volume equal to that of 25 Amazon Rivers! The NADW cruises the deep ocean, eventually meeting up with the AABW, the densest water mass on Earth. Figure 1 shows a schematic of the entire path of the Global Conveyor Belt that connects surface and bottom waters in all five oceans. In Figure 1 the blue lines represent deep currents and the red lines represent shallow currents.

Global Conveyor Movie

To better understand the global conveyor belt, click the Media Viewer button and choose **Global Conveyor** from the media list.

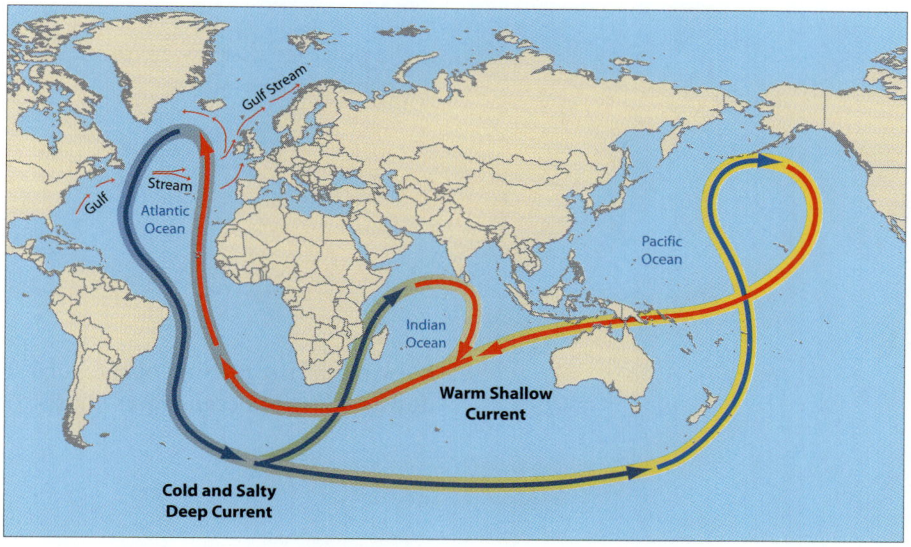

Figure 1. Simplified diagram of the Global Conveyor Belt.

1. Besides water, what else might density currents transport to the bottom of the ocean that is crucial to deep-sea life?

2. The average temperature of the ocean is 3.5 °C, whereas the average temperature of the ocean's *surface* is around 9 °C. Does this make sense? Explain.

Scientists speculate that 12,000 years ago the Global Conveyor Belt shut down, and that it could do so again in the future. They hypothesize that the density currents stopped forming in response to changes in salinity. Polar melting, changes in global winds and ocean-surface currents, and changes in precipitation patterns could all contribute to shutting down density currents. The climate and ocean currents are part of a very complex system, which makes them challenging to model precisely. Let us imagine a scenario in which a climate change results in a warmer Earth. Answer the questions below based on what you have learned in this exercise.

3. How would this climate change (a warmer Earth) influence the polar ice sheets that contain much of Earth's freshwater?

4. If the polar ice sheets were to melt, how might this influence the density of the ocean surface water?

5. How might the melting of the polar ice sheets influence the upwelling and downwelling of ocean-water masses?

6. What might be some environmental consequences if the Global Conveyor Belt were to shut down?

Unit 3
Ocean-Atmosphere Interactions

In this unit, you will

- ***Explore how the distribution of land masses affects global temperature patterns.***
- ***Discover how the ocean and atmosphere transfer energy from equatorial regions toward the poles.***
- ***Compare ocean and atmospheric conditions between normal, El Niño, and La Niña phases.***
- ***Investigate the effects of El Niño Southern Oscillation on climate in different parts of the world.***

NASA/GSFC

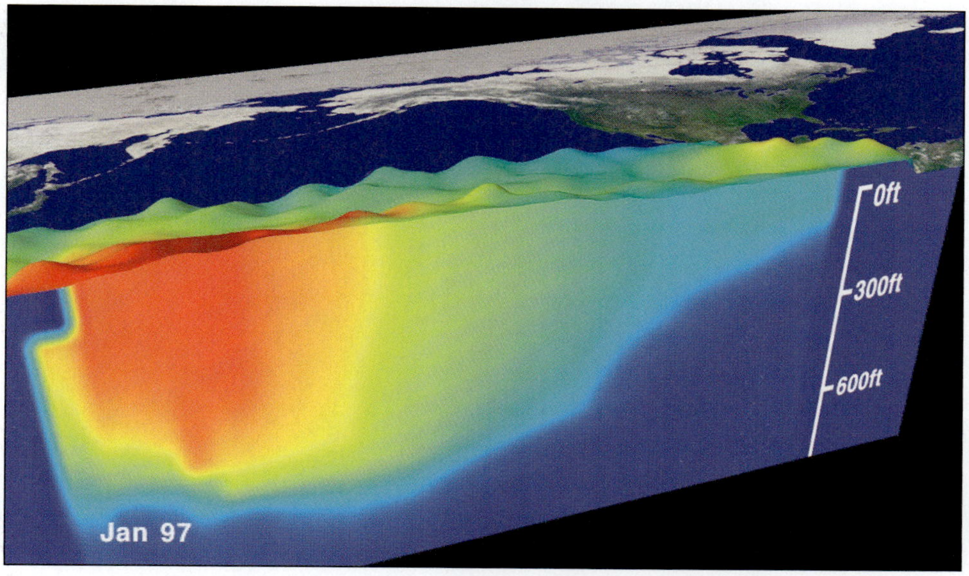

Cross section of the equatorial Pacific Ocean in January 1997. Warm temperatures are shown in red and orange, cold in blue. Ocean depths are greatly exaggerated.

Warm-up 3.1

Earth's extremes

The hottest surface air temperature ever recorded on Earth was 57.8 °C (136 °F), at El Azizia, Libya on September 13, 1922.

The coldest surface air temperature ever recorded on Earth was – 89 °C (– 128.6 °F), at Vostok, Antarctica on July 21, 1983.

Moderating global temperature

On the moon, daytime surface temperatures at the equator average a toasty 135 °C (275 °F), while the poles chill at about –225 °C (–373 °F), a range of 360 °C (648 °F)! On Earth, the difference between the hottest and coldest temperatures ever recorded is only 147 °C (265 °F). This is still a wide range of temperatures, but not beyond what humans can endure. What is it about the oceans and atmosphere that help moderate Earth's temperature?

1. List and describe three factors about the ocean and atmosphere that you think could moderate global temperatures.

   a.

   b.

   c.

Between 40° N and 35° S, Earth receives more radiation from the sun than it gives off, and most of it is stored in the oceans. At higher latitudes, more radiation is lost than received. If the oceans and atmosphere did not circulate and redistribute this energy, sea-surface temperatures would increase from the equator to the poles, forming bands of equal temperature parallel to the equator. Furthermore, average temperatures at the equator would be as much as 14 °C (25 °F) warmer and the poles would be as much as 25 °C (45 °F) colder than they actually are.

Fortunately, the oceans and atmosphere do circulate, moving energy from the equator toward the poles. Next, you will examine variations in ocean temperatures and make predictions about the effect they have on land.

Energy in the balance

- Launch ArcMap, and locate and open the **etoe_unit_3.mxd** file.

Refer to the tear-out Quick Reference Sheet located in the Introduction to this module for GIS definitions and instructions on how to perform tasks.

- In the Table of Contents, right-click the **Sea-Surface Temperatures** data frame and choose Activate.
- Expand the **Sea-Surface Temperatures** data frame.

JJA = June, July, and August

DJF = December, January, and February

The **DJF SST** layer shows average sea-surface temperatures during the months of December, January, and February. Note where the warmest temperatures are located.

Turn on the **JJA SST** layer.

The **JJA SST** layer shows average sea-surface temperatures during the months of June, July, and August. Watch for changes in the locations of the warmest sea-surface temperatures as you switch from the **DJF SST** to the **JJA SST** layer.

Seasonal changes in sea-surface temperature may be easier to see if you view a sequence of images covering a full year.

Click the Media Viewer button and view the **SST Movie** several times. Note that the center of the map in the movie is at a different longitude than the center of the map in ArcMap.

2. Describe how the location of the maximum sea-surface temperature region appears to change from December–February to June–August. (In which direction does it move? See Figure 1 at left for help.)

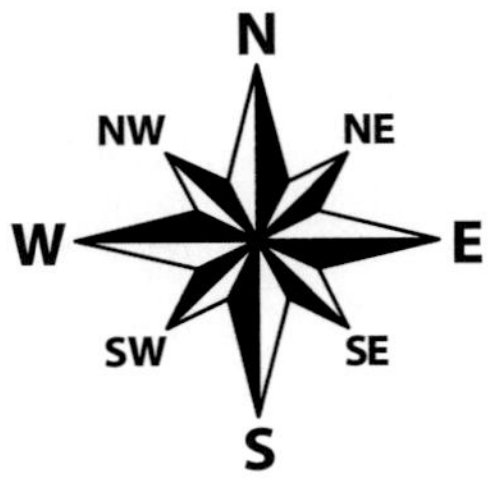

Figure 1. Compass directions.

Close the Media Viewer window.

Using both the **DJF SST** and **JJA SST** layers, look for areas where warmer water (reds) extends toward the poles. These areas represent warm currents flowing from the equator toward the poles. You can visualize these areas as tongues of water that are warmer than the surrounding waters at the same latitude.

Current directions

In general, the Coriolis effect causes currents to flow in a *clockwise* direction in the Northern Hemisphere and in a *counterclockwise* direction in the Southern Hemisphere.

3. Draw *solid* lines on Map 1 to show the locations of warm currents. Use arrowheads to indicate the direction of each current. The warm current off the east coast of South America has been drawn for you as an example. (See **Current directions** at left.)

Map 1 — Global ocean currents

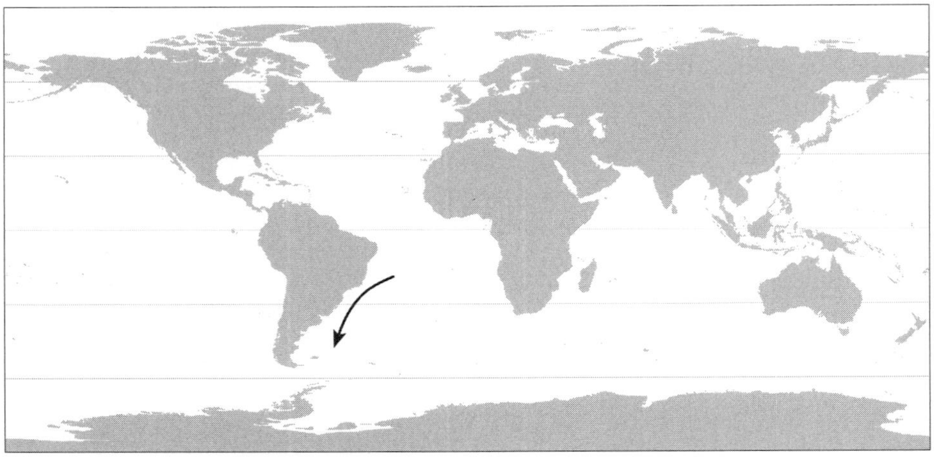

Look for areas where cooler water (blues) extends toward the equator. These areas represent cool currents flowing from the poles toward the equator.

4. Draw *dashed* lines on Map 1 to show the locations of cool currents. Use arrowheads to indicate the direction of each current.

Someone once said *"The coldest winter I ever spent was a summer in San Francisco."*

Who said that?

This quote is often attributed to Mark Twain, but there is no evidence that he ever said or wrote it.

5. Using the **JJA SST** layer and Map 1, examine the currents near California during the summer. How might these patterns affect temperatures in San Francisco to support this quote?

6. Describe how you think life on Earth would be different if there were no ocean circulation and the equator were 14 °C hotter and the poles 25 °C colder than they are today.

In the next investigation, you will explore how the distribution of land and ocean affects climate.

💻 Quit ArcMap and do not save changes.

Investigation 3.2

Heat energy

Thermal inertia is similar to other terms that may be more familiar, such as *heat capacity*, or *specific heat*. These refer to the amount of energy required to raise the temperature of an object by 1 °C.

A tale of two hemispheres

Thermal inertia, a measure of how difficult it is to change the temperature of a material, has a strong effect on how Earth stores and releases solar energy. Water has a high thermal inertia, typically four times that of soil or air. Thus, the upper layer of the ocean stores much more solar energy and it changes temperature much more slowly than does the land or the atmosphere. For this reason, the ocean is cooler during the day and warmer during the night relative to the atmosphere or land.

1. Based on your current understanding, circle the words that you think best complete the following statement:

    The land heats up more quickly / slowly and cools off more quickly / slowly than the ocean, so the range of temperatures over the land should be greater / smaller than over the ocean.

In this investigation, you will explore temperature patterns in the atmosphere and oceans around the world to better understand the role the ocean plays in moderating global climate.

Seasonal temperature variation

Launch ArcMap, and locate and open the **etoe_unit_3.mxd** file.

Refer to the tear-out Quick Reference Sheet located in the Introduction to this module for GIS definitions and instructions on how to perform tasks.

In the Table of Contents, right-click the **Global Energy** data frame and choose Activate.

Expand the **Global Energy** data frame.

The **Seasonal Avg Air Temp Range** layer shows the difference in the average surface air temperatures between June, July, and August (JJA) and December, January, and February (DJF). Warm summers and cold winters result in a wide temperature range, whereas mild summer and winter temperatures or consistently cold or hot temperatures result in a narrower range.

Look at the distribution of the seasonal temperature ranges in the Northern and Southern Hemispheres.

2. In general, over which type of surface (land masses or oceans) does the seasonal temperature variation appear to be the
    a. largest?

    b. smallest?

    c. Why do you think this pattern occurs?

- Turn off and collapse the **Seasonal Avg Air Temp Range** layer.

To better understand these temperature range variations, you will examine the distribution of land and ocean in the Northern and Southern Hemispheres.

The distribution of land and ocean

Over 70 percent of the world's surface is covered by ocean, with land covering the remaining 29 percent. However, the land is not distributed evenly over the globe.

- Turn on the **Land and Ocean** layer.
- Select the **Land and Ocean** layer.

Visual estimation

To learn more about visually estimating the percentage of coverage, see the section on *Estimating percent areas* in the Introduction to this guide.

3. Visually estimate the percentage of ocean and land in each hemisphere and enter your estimates in Table 1.

Table 1 — Estimated percentages of land and ocean by hemisphere

Hemisphere	% Land	% Ocean	Total
Northern			100%
Southern			100%

You will evaluate your estimate by calculating the actual percentage of each hemisphere that is covered by ocean.

- Click the Select By Attributes button.
- To select the Northern Hemisphere oceans, query the **Land and Ocean** layer for **("Hemisphere" = 'Northern') and ("Surface" = 'Ocean')** as shown in steps 1–6. The query will actually read:

("HEMISPHERE" = 'Northern') AND ("SURFACE" = 'Ocean')

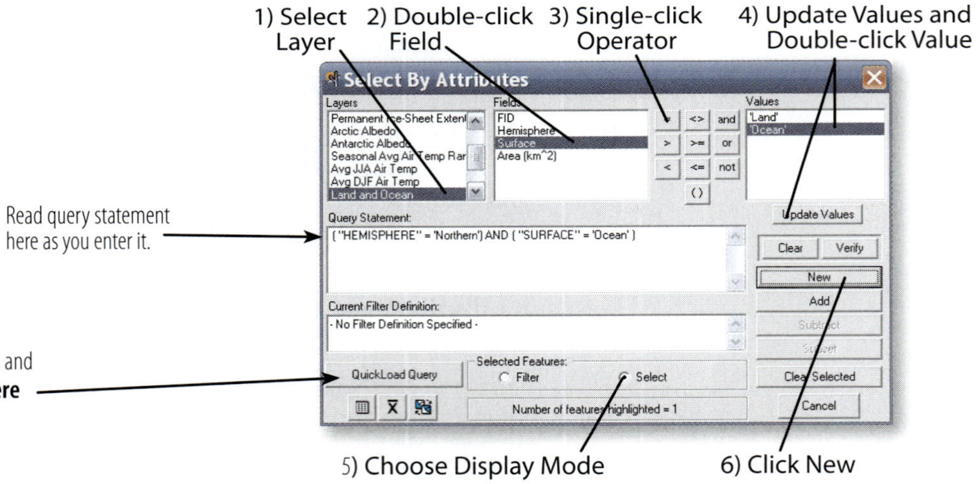

QuickLoad Query

- Click the QuickLoad Query button and select the **Northern Hemisphere Ocean** query.
- Click **OK**.
- Click **New**.

- If you have difficulty entering the query statement correctly, refer to the **QuickLoad Query** described at left.
- Do not close the Select By Attributes window.

The outline of the oceans in the Northern Hemisphere should be highlighted. Next you will calculate statistics for the selected data.

- Click the Statistics button in the Select By Attributes window.

In the Statistics window, calculate statistics for **only selected features** of the **Land and Ocean** layer, using the **Area (km^2)** field.

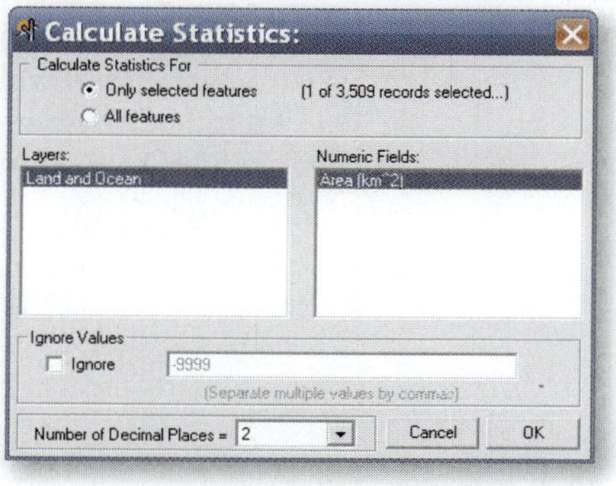

Click **OK**.

The total area of the Northern Hemisphere oceans is reported as the **Total** in the Statistics window.

Rounding

To learn more about rounding, see the section on *Rounding* in the Introduction to this guide.

4. Record the area of the oceans in the Northern Hemisphere in Table 2. (Land area is already provided for you.) Round all values to the nearest million km^2.

Table 2 — Ocean and land area by hemisphere

Hemisphere	Ocean area *million km²*	Land area *million km²*	Total area *million km²*	% Ocean area
Northern		101		
Southern		47		

Close the Statistics window.

Click **Clear** in the Select By Attributes window.

Repeat the Select By Attributes operation to find the total area for the Southern Hemisphere's oceans. Query the **Land and Ocean** layer for **("Hemisphere" = 'Southern') and ("Surface" = 'Ocean')**. The query will actually read:

("HEMISPHERE" = 'Southern') AND ("SURFACE" = 'Ocean')

QuickLoad Query

If you have difficulty entering the query statement correctly:

- Click the QuickLoad Query button and select the **Southern Hemisphere Ocean** query.
- Click **OK** .
- Click **New**.

If you have difficulty entering the query statement correctly, refer to the **QuickLoad Query** described at left.

Repeat the Statistics process to calculate statistics for **only selected features** of the **Land and Ocean** layer, using the **Area (km^2)** field.

The total area of the Southern Hemisphere oceans is reported as the **Total** in the Statistics window.

5. Record the area of the oceans in the Southern Hemisphere in Table 2. Round all values to the nearest million km^2.

Calculate the total area of each hemisphere by adding the Ocean Area and Land Area and record your results in Table 2. (Hint: The total areas of the Northern and Southern Hemispheres should be nearly equal.)

Calculating % Ocean Area

$$\%\ \text{Ocean Area} = \frac{\text{Ocean Area}}{\text{Total Area}} \times 100$$

Calculate the % Ocean Area for each hemisphere (see sidebar for help). Round your results to the nearest 0.1 percent and record them in Table 2.

Close the Statistics and Select By Attributes windows.

Remember that land masses respond more quickly to changes in air temperature and solar radiation, so a land mass will warm or cool more quickly than the ocean.

6. Using this information and the measurements recorded in Table 2, determine in which hemisphere you would expect to find the warmest temperatures, and in which hemisphere you would expect to find the coldest temperatures?
   a. Warmest

   b. Coldest

7. How do you think the Earth's daily temperature range would be affected if there were no oceans?

- Click the Clear Selected Features button.
- Turn off the **Land and Ocean** layer.

In the next section, you will determine in which hemisphere the warmest and coldest air temperatures are found, then examine worldwide surface air temperatures during the warmest and coldest seasons.

Air temperature

- Turn on and expand the **Avg JJA Air Temp** layer.

The **Avg JJA Air Temp** layer shows the average surface air temperatures for the months of June, July, and August, corresponding to Northern Hemisphere summer and Southern Hemisphere winter. Reds represent higher temperatures and blues represent lower temperatures.

JJA = June, July, and August
DJF = December, January, and February

- Examine the **Avg JJA Air Temp** legend and find areas with the highest and lowest average air temperatures at this time of year.

You will now summarize the Northern and Southern Hemisphere minimum and maximum temperatures for the **Avg JJA Air Temp** layer.

- Click the Summarize button Σ.
- In the Summarize window, select **Avg JJA Air Temp** as the **feature layer**.
- Select **Hemisphere** as the **field to summarize** in the drop-down menu.
- Double-click **Temp (C)** to display the statistics options, check **Minimum** and **Maximum**, and click **OK**.

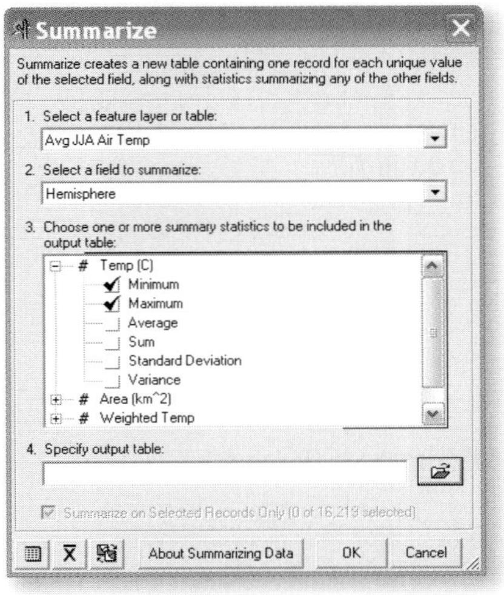

In the summary table, the **Minimum_Temp_C_** and **Maximum_Temp_C_** fields give the minimum and maximum air temperature for the Northern and Southern Hemispheres for the months of June, July, and August.

8. Using the resulting summary table, record these temperature data in the first row of Table 3. Round to the nearest 1 °C.

Table 3 — Seasonal temperature extremes

Time of year	Northern Hemisphere			Southern Hemisphere		
	Minimum temp. °C	**Maximum temp.** °C	**Range** *(Max. – Min.)* °C	**Minimum temp.** °C	**Maximum temp.** °C	**Range** *(Max. – Min.)* °C
June – August						
December – February						

Close the Summary Table window when you are finished.

Turn off and collapse the **Avg JJA Air Temp** layer.

Turn on and expand the **Avg DJF Air Temp** layer.

This layer shows average air temperatures for December - February, corresponding to Northern Hemisphere winter and Southern Hemisphere summer. You will now do the same summarize operation for the Northern and Southern Hemisphere minimum and maximum temperatures for the **Avg DJF Air Temp** layer.

Click the Summarize button [Σ].

In the Summarize window, select **Avg DJF Air Temp** as the **feature layer.**

Select **Hemisphere** as the **field to summarize** in the drop-down menu.

Double-click **Temp (C)** to display the statistics options, check **Minimum** and **Maximum**, and click **OK**.

In the summary table, the **Minimum_Temp_C_** and **Maximum_Temp_C_** fields give the air temperature for both hemispheres for December-February.

9. Using the resulting summary table, record these temperature data in the second row of Table 3. Round to the nearest 1 °C.

10. Calculate the temperature ranges for June–August and for December–February for the Northern and Southern Hemispheres and record them in Table 3.

11. Circle the highest **Maximum Temp. (°C)** and the lowest **Minimum Temp. (°C)** in Table 3, regardless of hemisphere. Record the difference in temperature between those two numbers here.

12. Using the information in the left margin about Earth's seasons and your knowledge about thermal inertia, explain why the global temperature range is so much larger in June–August than in December–February. (Hint: Remember the distribution of land and water in the Northern and Southern Hemispheres.)

Earth's seasons are...

- caused by the tilt of Earth's axis.
- opposite in the Northern and Southern Hemispheres.

Dates (typical)	Hemisphere	
	Northern	**Southern**
Dec 21 – Mar 20	Winter	Summer
Mar 20 – Jun 21	Spring	Fall
Jun 21 – Sep 22	Summer	Winter
Sep 22–Dec 21	Fall	Spring

- Close the Summary Table window.
- Turn off and collapse the **Avg DJF Air Temp** layer.

Temperature range and ocean proximity

You have explored how air temperature varies by hemisphere and how the amount of ocean versus land can affect that variation. Next, you will investigate the role oceans play in moderating climate by examining how temperatures change as you move inland, away from the ocean.

- Turn on and expand the **Temperature Range** layer. (Note: This is not the same as the **Seasonal Avg Air Temp Range** layer.)

The **Temperature Range** layer shows the difference between the average summer and winter temperatures for over 6000 locations around the world. Red represents large seasonal temperature differences and blue represents small seasonal temperature differences.

- Click the Summarize button **Σ**.
- In the Summarize window, select **Temperature Range** as the **feature layer**.
- Select **Coastal Proximity** as the **field to summarize** in the drop-down menu**.**
- Double-click **Temp Range (C)** to display the statistics options, check **Average**, and click **OK**.

The fourth column of the summary table, labeled **Average_SEASONAL_D,** contains the average temperature range for each of the coastal proximity categories.

13. Record the average temperature range for each of the four proximity categories in Table 4. Round to the nearest 1 °C. (Caution: The categories in the summary table are NOT listed in the same order as in Table 4. Be sure you are entering the correct temperature range value for each category.)

Table 4 — Summary of average temperature range by coastal proximity

Increasing distance from ocean ↓

Coastal proximity category	Average temperature range (°C)
Coastal (0 – 80 km)	11
Near-coastal (80 – 160 km)	
Mid-coastal (160 – 800 km)	
Continent interior (over 800 km)	

14. On Graph 1, make a bar graph of the data located in Table 4.

Graph 1 — Temperature range versus proximity to the coast

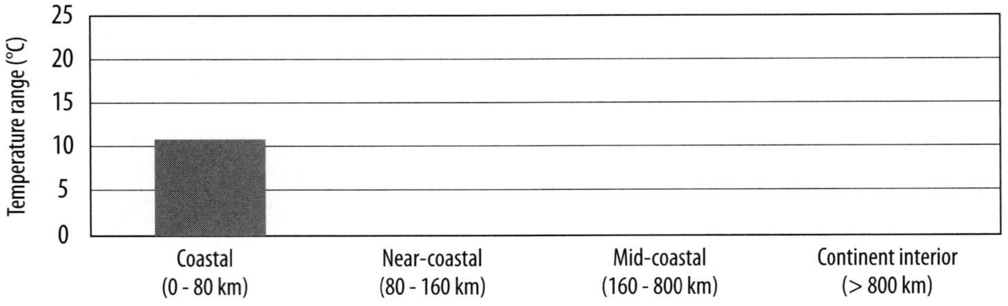

Close the Summary Table window.

15. Study Graph 1, then describe the ocean's influence on seasonal temperature range as distance inland from the coast increases.

Turn off and collapse the **Temperature Range** layer.

Polar albedo

energy balance — the balance between incoming solar radiation and outgoing radiation.

Although the oceans are very effective at storing solar energy, the polar ice sheets are excellent at reflecting solar energy. Reflectance, also called *surface albedo*, is an important component of Earth's global energy balance. Many of us have experienced the effects of albedo through the clothes that we wear. White clothes reflect solar radiation and help us stay cooler, whereas dark or black clothes absorb solar energy and make us feel warmer. This is true for Earth's surface as well. The north and south polar regions have different albedos, which affect their temperatures.

16. Review the temperature data for winter in each hemisphere (DJF in Northern Hemisphere and JJA in Southern Hemisphere) in Table 3 (page 91). Predict which polar region you think will have the highest albedo (is most reflective), and circle the words in the statement below that summarize your prediction.

    The coldest winter temperatures occur in the Northern / Southern Hemisphere, which suggests that the greatest reflectance or albedo should be found in the northern / southern polar region.

Turn on the **Permanent Ice-Sheet Extent** layer.

Turn on and expand the **Antarctic Albedo** layer.

Turn on and expand the **Arctic Albedo** layer.

Select **Arctic Albedo** layer.

In these layers, darker shades of brown represent lower albedos (less reflective surfaces) and lighter shades represent higher albedos (more reflective surfaces).

As you examine the Antarctic and Arctic albedo layers, note that some of the darker albedo values around both polar caps represent water during times when no sea ice is present. The permanent ice sheets of Greenland and Antarctica are clearly visible, whereas the approximate extent of the permanent ice sheet over the North Pole is identified by the dark blue line.

17. Which pole appears to have the higher albedo? (Darker browns = lower albedo and lower reflectivity; lighter browns and white = higher albedo and higher reflectivity.)

Now you will check your prediction by calculating the average albedo of each polar region.

Click the Select By Attributes button.

QuickLoad Query

- Click the QuickLoad Query button and load the **Arctic Ocean** query.
- Click **OK** .
- Click **New**.

No features found?

If the **Number of features selected** at the bottom of the Select By Attributes window is 0, redo the query. Be sure to spell *Arctic* correctly.

- To calculate the average Arctic Albedo, query the **Arctic Albedo** layer for **("Location" = 'Arctic Ocean').** The query will actually read:

  ("LOCATION" = 'Arctic Ocean')
- If you have difficulty entering the query statement correctly, refer to the **QuickLoad Query** described at left.
- Click **New**. Do not close the Select By Attributes window.

The Arctic Ocean region will be highlighted. Next you will calculate statistics for the selected data.

- Click the Statistics button [X̄] in the Select By Attributes window.
- In the Statistics window, calculate statistics for **only selected features** of the **Arctic Albedo** layer, using the **Reflectance (%)** field.
- Click **OK**.

The average reflectance of the Arctic Ocean region is reported in the Statistics window as the **Mean**.

18. What is the average reflectance (%) of the Arctic Ocean region? Round to the nearest 1 percent.

- Close the Statistics window.
- Click **Clear** in the Select By Attributes window to clear the previous query.
- Select the **Antarctic Albedo** layer.
- Repeat the Select By Attributes operation, and query the **Antarctic Albedo** layer for **("Location" = 'Antarctica').** The query will actually read:

  ("LOCATION" = 'Antarctica')

QuickLoad Query

- Click the QuickLoad Query button and load the **Antarctica** query.
- Click **OK** .
- Click **New**.

- If you have difficulty entering the query statement correctly, refer to the **QuickLoad Query** described at left.
- Click **New**.

The Antarctic Ocean region will be highlighted. Next you will calculate statistics for the selected data.

- Click the Statistics button [X̄] in the Select By Attributes window.
- In the Statistics window, calculate statistics for **only selected features** of the **Antarctic Albedo** layer, using the **Reflectance (%)** field.
- Click **OK**.

The average reflectance of Antarctica is reported in the Statistics window as the **Mean**.

19. What is the average albedo percentage of the Antarctic region? Round to the nearest 1 percent.

- Close the Statistics and Select By Attributes windows.

20. How does the albedo you measured for each polar region compare to your prediction?

The Southern Hemisphere is covered by land and ice in the polar region, whereas the Northern Hemisphere is covered by water and ice.

21. How might the coverage of land and ice in the Southern Hemisphere and water and ice in the Northern Hemisphere influence the albedo in each Hemisphere?

Click the Clear Selected Features button.

Quit ArcMap and do not save changes.

In the next investigation, you will learn more about Earth's global energy balance and its impact on climate.

Reading 3.3

Climate oscillations

Our climate is the product of interactions between the atmosphere, hydrosphere (Earth's liquid water), cryosphere (glaciers and ice sheets), geosphere (solid Earth), and biosphere (living organisms) that are driven by solar energy. The ocean and the atmosphere have the greatest influence on our climate. The oceans have a strong moderating effect on global temperatures, while the atmosphere reflects harmful solar radiation and traps heat and moisture near the surface. However, all of Earth's systems are linked together, and changes to any one system can have a substantial impact on the others. To fully understand the dynamics of our climate, we must examine the global energy balance and the transfer of energy among these systems.

Global energy balance

infrared radiation (IR) — invisible form of electromagnetic radiation that is sensed as heat.

latent heat — heat energy released or absorbed during a change of phase.

The Sun is the primary source of energy controlling temperatures on and near Earth's surface. When solar energy reaches Earth's atmosphere, approximately 30 percent is reflected back into space (~25% from clouds and ~5% from the ground); 51 percent is transmitted to the ground or ocean surface and absorbed; and 19 percent is absorbed by gases in the atmosphere (Figure 1 below). Of the 51 percent of the solar energy absorbed by the surface, 7 percent is used to raise the temperature of the atmosphere (sensible heat flux), 21 percent is radiated back to the atmosphere as **infrared radiation** (IR), and another 23 percent is used to evaporate water (**latent heat** flux) from the ocean and soils. Ultimately, all incoming solar energy is radiated back to space—otherwise, Earth would steadily heat up.

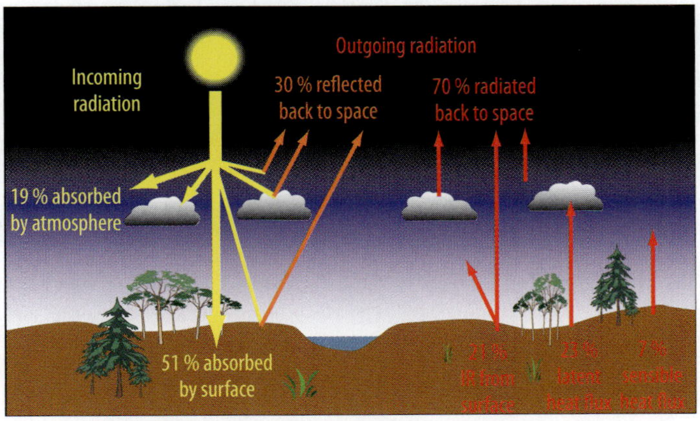

Figure 1. Annual mean global energy balance.

The total amount of solar energy reaching Earth each year is constant, but it is not distributed evenly across the globe. This uneven distribution of energy creates daily and seasonal temperature differences that drive winds, currents, precipitation, and evaporation (Figure 2 on the following page).

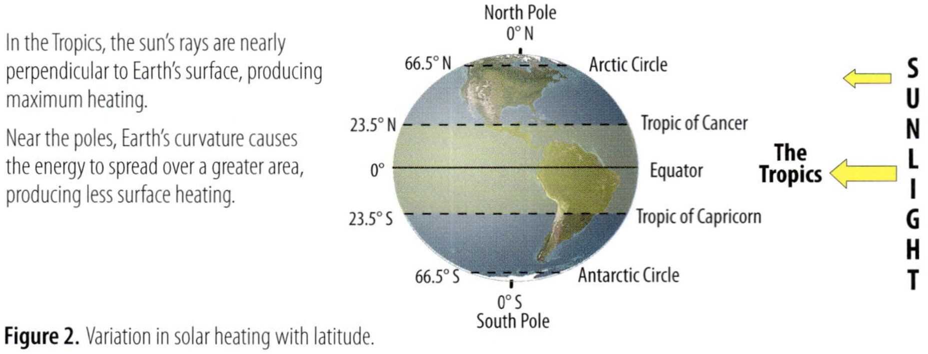

Figure 2. Variation in solar heating with latitude.

Storing energy

To understand how solar radiation affects large-scale processes such as winds, currents, and climate, you need to know a few things about water and heat energy. Water can exist in three **states**—solid, liquid, or vapor (Figure 3). When water changes states, it absorbs or releases more heat (called latent heat) than do most other substances.

heat capacity—the amount of energy required to raise the temperature of one unit of mass of a substance by 1 °C.

Liquid water can absorb more energy than any other liquid except ammonia, because it has a high **heat capacity**. Raising the temperature of water requires a lot of energy, so it is very difficult to change the temperature of the ocean even a small amount. This resistance to change is called **thermal inertia**. However, even small changes to the energy content of the ocean can have considerable effects on global climate. In contrast, the heat capacity and thermal inertia of the atmosphere are much lower, making it easier to change the temperature of the atmosphere.

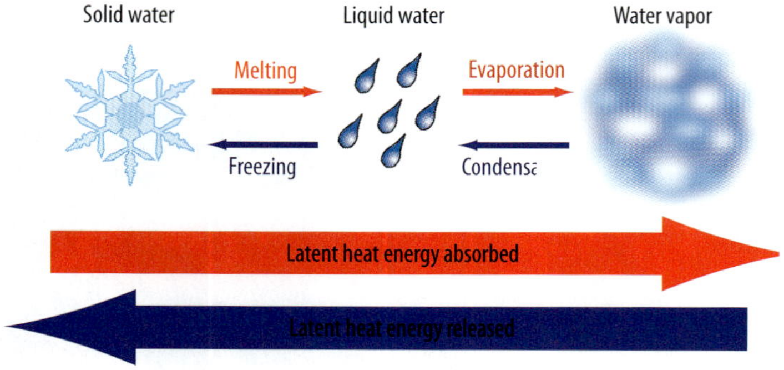

Figure 3. As ice melts or water evaporates, it absorbs latent heat; and as water vapor condenses or water freezes, it releases latent heat. The transition between liquid water and water vapor involves about seven times as much heat energy as the transition between liquid water and ice.

At the equator, where solar energy is highest, about 75 percent of Earth's surface is covered by water. Water's higher heat capacity allows the ocean to absorb and retain more solar energy than the land or atmosphere. In fact, the upper few hundred meters of ocean stores approximately 30 times more heat than the entire atmosphere. Without some type of circulation in the ocean and atmosphere, the equator would be 14 °C (25 °F) warmer on average than now, and the North Pole would be 25 °C (45 °F) colder. Fortunately, temperature imbalances drive atmospheric and oceanic circulation, which redistribute energy over Earth's surface.

1. How might global temperatures change if 75 percent of the area near the equator were covered with land instead of water? Explain.

Reflecting energy

Earth's ability to store solar energy is critical for regulating global temperatures. Equally important is its ability to reflect solar energy. The reflectivity of a material—the ratio of the amount of light reflected by an object to the amount of light that falls on it—is called its **albedo**. Albedo is determined principally by a material's texture and color: Rougher, darker surfaces tend to have lower albedos; whereas smooth, bright surfaces reflect more efficiently and have higher albedos. In addition, the angle at which the sun's rays strike the surface affects the albedo.

Albedo of surface types at 45° N latitude

Surface Type	Albedo (%)
dense swampland	9 – 14
deciduous trees	13
grassy field	20
barren field	5 – 40*
farmland	15
desert or large beach	25*

* Depending on color of surface.

The albedos of various surface materials range widely, from extremely reflective to highly absorptive. For instance, different states of water—as ice, liquid water, and water droplets—have very different albedos. Ice and certain types of clouds have a high albedo and reflect 60 – 90 percent of the sun's energy. The oceans, on the other hand, generally have a low albedo (10%), which means that they absorb 90 percent of the sun's energy that strikes them. The albedo of the biosphere, meanwhile, varies according to the type of land cover; vegetated land has a much lower albedo than barren land. Thus, tropical rainforests absorb more solar energy than deserts.

The average albedo for the whole Earth is about 30 percent. This is not a permanent characteristic; any significant alteration of Earth's surface, such as melting of the ice sheets, desertification, or burning of tropical rain forests, can trigger changes in the amount of solar radiation absorbed or reflected. A change in albedo is significant, because global albedo is a key determinant of conditions on Earth. For example, if the ice sheets melted, the global albedo would decrease. Earth's surface would then absorb more energy, and the temperature of the atmosphere would rise. Evidence in the geologic record suggests this happened during the Cretaceous Period (144 to 65 million years ago). At that time, there was little or no snow and ice cover, even at the poles, and global temperatures were at least 8 – 10 °C warmer than today. Such global warming hinders the formation of sea ice and disrupts the development of cold, dense currents that sink to the depths of the ocean. In addition, warmer temperatures melt the ice sheets. This, in turn, increases the volume of the oceans, raises global sea levels, and triggers changes in Earth's energy balance.

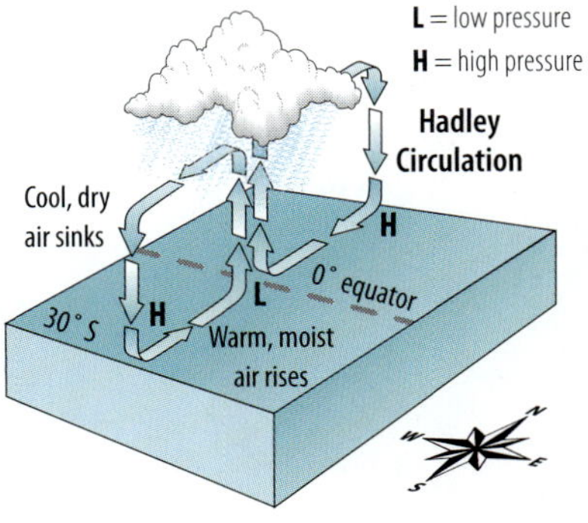

Figure 4. Simple convection cells between the equator and 30° N and 30° S move energy from the equator toward the poles. This process is called **Hadley circulation**.

Circulating energy with winds and currents

The uneven distribution of solar energy produces circulation in the ocean and atmosphere that moderates temperatures across the globe. Near the equator, evaporation adds large amounts of water vapor and latent heat to the atmosphere (Figure 4). Hot, moist air rises from the ocean surface and cools, causing some of the water vapor to condense and fall back to Earth as rain. As water vapor condenses, it releases latent heat. This generates temperature differences and winds higher in the atmosphere, which transport the remaining water vapor and latent heat toward the poles. As the air moves toward the poles, it continues to cool, becomes denser, and descends, flowing back towards the equator to complete the cycle.

Ocean currents contribute to the heat transfer between the equator and the poles by moving warm water toward the poles and cold water toward the equator. Warm surface currents at the equator are driven westward by the winds and deflect toward the poles as they near continents. As these warm currents flow toward the poles, they transfer heat to the atmosphere, warming the atmosphere and cooling the currents. The cooled surface currents then flow back toward the equator to be warmed again.

Deep-water density currents also play an important role in redistributing energy. Near the poles, cold, salty waters sink to great depths and flow slowly toward the opposite pole. These currents eventually rise to the surface where the water is reheated to begin the cycle again.

The Southern Oscillation

oscillation — back and forth pattern of change between one state or direction and its opposite.

The ocean's high thermal inertia (slow response to change) helps stabilize Earth's climate. However, this stability is not a constant. Over the past century, scientists have observed **oscillations** in sea-surface temperature and air pressure that occur over years and decades. Localized ocean-temperature increases of only 2 – 3 °C have been linked to extended periods of drought and flooding globally. Heat transfer between the ocean and atmosphere, which causes changes in surface air pressure and winds, drives the oscillations. The best known of these is the **Southern Oscillation**, a seesaw shift in surface air pressure driven by changes in sea-surface temperature between the eastern and western Pacific Ocean in the Southern Hemisphere.

What drives the Southern Oscillation?

temperature gradient — change in temperature over a distance.

Strong evaporation and convection at the ocean's surface drive the Southern Oscillation. In a normal year, the average sea-surface temperature is warmer in the western Pacific than in the eastern Pacific (Figure 5). This **temperature gradient** causes air to circulate parallel to the equator in a pattern called the **Walker Circulation**. Warm waters of the western Pacific produce high levels

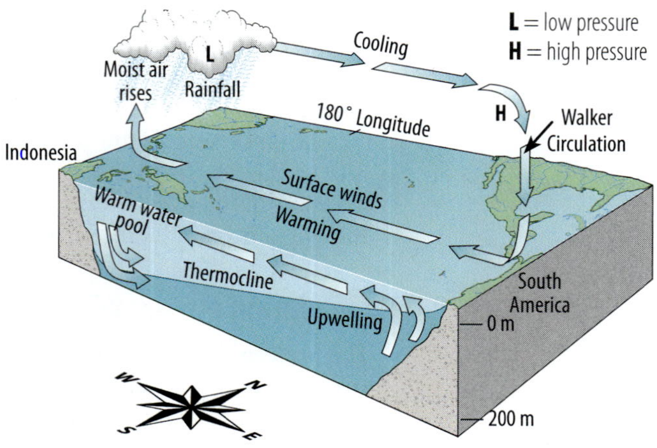

Figure 5. Normal conditions in the equatorial Pacific Ocean.

of evaporation and increased precipitation. The warm, rising air creates low pressure at the surface. As surface winds blow into this low-pressure center, they pile up water to form a small mound in the western Pacific. As a result, the layer of warm surface water is deeper in the western Pacific than in the east. This forces the deeper and colder waters in the western Pacific to flow eastward and up toward the surface, producing cooler sea-surface temperatures off the coast of South America. The rising, moist air in the western Pacific fuels the heavy monsoon rains of Southeast Asia, and the sinking dry air in the eastern Pacific creates the coastal deserts of South America.

2. Circle the word in each pair that correctly completes the sentence:

   "Warmer / Cooler surface temperatures cause increased evaporation and upward convection of air masses, which leads to higher / lower air pressure near the surface."

Changes to the Southern Oscillation — El Niño and La Niña

phase — stage in a periodic process or phenomenon.

About every three to eight years, the Southern Oscillation strengthens or weakens and we experience either El Niño or its sister **phase**, La Niña. During an El Niño phase (Figure 6), a 1–2 °C warming of the water in the eastern Pacific reduces the sea-surface temperature gradient from east to west. The reduced temperature gradient weakens the east to west trade winds in the South Pacific. This allows the mound of warm water piled up in the western Pacific to spread out and move eastward. The warm surface waters in the east and high air pressures prevent the typical upwelling of cold, nutrient-rich water off the west coast of South America.

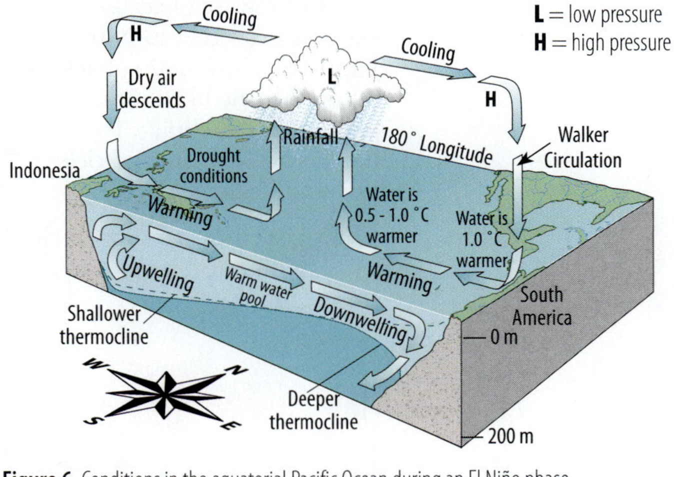

Figure 6. Conditions in the equatorial Pacific Ocean during an El Niño phase.

Although the El Niño Southern Oscillation occurs in the tropical Pacific Ocean, its impact on climate is observed globally. Variations in rainfall are significant, with reports of droughts in regions of Indonesia and Australia that are normally drenched by monsoons, and storms and flooding in Ecuador and parts of the United States. In addition to climate changes, El Niño phases can have devastating effects on fisheries, agriculture, and outbreaks of fire and disease. For example, in most years colder nutrient-rich water from the deeper ocean is drawn to the surface near the coast of Peru and Ecuador (upwelling). This produces abundant plankton, the food source of the anchovy, and the anchovy harvest increases. However, when upwelling weakens in an El Niño phase and warmer low-nutrient water spreads along the coast, the anchovy harvest plummets.

An El Niño phase that produces warming of the water in the eastern Pacific basin is counteracted in other years by a similar cooling phase known as La Niña. During a La Niña phase, trade winds are anomalously stronger than normal, and the sea-surface temperature gradient increases (Figure 7). This causes greater upwelling of cold water in the eastern Pacific Ocean and increased warming and evaporation in the western Pacific Ocean.

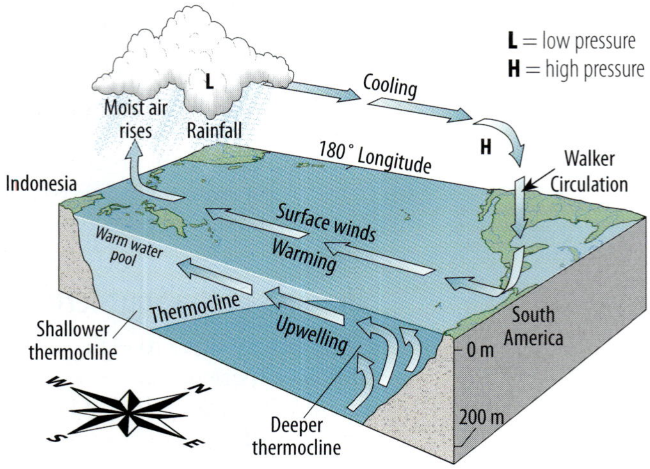

Figure 7. Conditions in the equatorial Pacific Ocean during a La Niña phase.

These oscillations and the conditions that produce them are the focus of extensive research because their climate impacts can be quite severe. A recent study suggests that El Niño can be triggered by volcanic eruptions. Climate and eruption records dating back to the 17th century suggest that a large eruption can double the likelihood of an El Niño phase the following winter. Volcanoes might alter the climate by spewing great quantities of dust and greenhouse gases into the air. The dust particles reflect sunlight and alter the amount of solar heat reaching Earth's surface.

3. List three ways in which El Niño might affect your everyday life. Consider both direct and indirect effects.

   a.

   b.

   c.

Other climate oscillations

The El Niño Southern Oscillation is the most studied and best understood ocean-atmosphere interaction. In recent decades, other patterns have been identified around the globe, including the Pacific Decadal Oscillation and the Atlantic Multidecadal Oscillation. The Pacific Decadal Oscillation is the primary factor determining variations in monthly sea-surface temperature in the North Pacific (poleward of 20° N). The Pacific Decadal Oscillation has a distinct pattern of long-term sea-surface temperature oscillations in the North Pacific. However, it occurs over 20–30 years, whereas typical El Niño Southern Oscillation phases last for only 6–18 months.

A similar pattern of sea-surface temperature oscillations in the North Atlantic Ocean—the Atlantic Multidecadal Oscillation—occurs over a period of 50–80 years. Scientists believe that the Atlantic Multidecadal Oscillation and Pacific Decadal Oscillation are linked to the major droughts of the past 100 years in North America including the Dust Bowl in the 1930s, extended droughts in the 1950s and 1970s, and the drought in the southwestern United States that started in 1998. Understanding these ocean-atmosphere phenomena is critical to understanding global climate variability and improving climate-forecasting capabilities. For at least the past 400 years, people living around the equatorial Pacific Ocean have noted that in some years the surface waters of the central and eastern Pacific Ocean are warmer than usual. Because this change usually begins in December, the phenomenon was named El Niño—Spanish for "the boy child" (referring to the Christ child). Over time, scientists have come to realize that El Niño is not an isolated phenomenon. Subsequent years sometimes show a pattern of unusually cool surface waters in this region, an event now referred to as the La Niña ("the girl child") phase. Both phases can have dramatic effects on global climate. Too wet, too dry, too hot, or too cold—during an El Niño or La Niña phase, chances are good that the weather will be unusual.

Investigation 3.4

El Niño and La Niña

Normal phase

phase — stage in a periodic process or phenomenon.

In the first part of this investigation, you will examine typical ocean and atmospheric circulation patterns in the Pacific Ocean under normal *phase* conditions. Later, you will compare these patterns to El Niño and La Niña phases to see how small changes can disrupt local circulation patterns and global weather conditions.

- Launch ArcMap, and locate and open the **etoe_unit_3.mxd** file.

Refer to the tear-out Quick Reference Sheet located in the Introduction to this module for GIS definitions and instructions on how to perform tasks.

- In the Table of Contents, right-click the **Normal Phase** data frame and choose Activate.
- Expand the **Normal Phase** data frame.

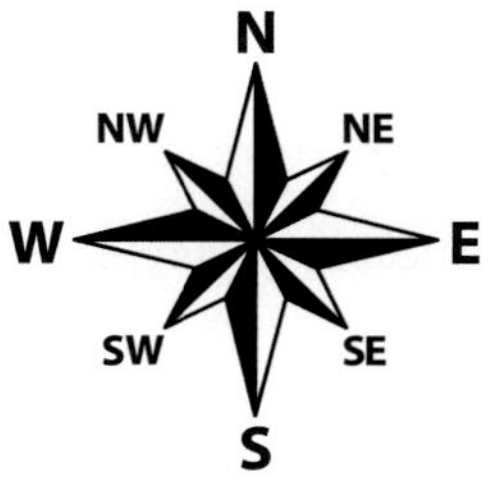

Figure 1. Compass directions.

The **Wind Speed** layer shows the average speeds and directions of surface winds around the globe. Arrows point in the average wind direction (Figure 1), and the color of each arrow represents the average wind speed at that location. Notice how, in this layer as well as many of the others, most characteristics of the ocean and atmosphere change more with latitude than with longitude.

- Click the QuickLoad button.
- Select **Spatial Bookmarks**, choose **Pacific Ocean**, and click **OK**.

1. Examine the surface winds over the Pacific Ocean between the equator and 30° latitude in both hemispheres, and describe any differences you observe
   a. *between* the Northern and Southern Hemisphere.

   b. from east to west in the Northern Hemisphere.

   c. from east to west in the Southern Hemisphere.

The winds blowing from east to west near the equator are called the *trade winds*, named for their importance in pushing the sailing ships that traveled between Europe and the Americas in the 17th and 18th centuries.

- Turn off the **Wind Speed** layer.
- Turn on and expand the **Surface Currents** layer.

The **Surface Currents** layer shows the average directions of the world's ocean surface currents. Note the directions of the ocean currents between 30° N and 30° S.

2. Between 30° N and 30° S, in what direction do most surface currents flow? How does this compare to the average direction of the trade winds? (Hint: Trade winds are defined on the previous page.)

Next, you will look at the average sea-surface temperatures (SST) across the Pacific Ocean.

- Turn off and collapse the **Surface Currents** layer.
- Turn on and expand the **Average Sea-Surface Temperature** layer.

Examine the sea-surface temperatures across the Pacific Ocean, paying particular attention to the region between 30° N and 30° S.

3. Which part of the equatorial Pacific Ocean appears to have the warmer surface temperature, the eastern Pacific or the western Pacific?

Sea not-so-level

You may have been taught that sea level is a flat surface that extends all over the ocean, but in fact the ocean surface has many small hills and valleys that vary from mean sea level by tens to hundreds of centimeters. These are caused by temperature and pressure differences, tides, gravity, and the effects of wind blowing across the ocean surface.

Water expands as it warms and contracts as it cools. This phenomenon is called *thermal expansion*.

4. Based on the average surface temperature alone, what effect do you think this expanding warmer water would have on sea level in this part of the Pacific? (Hint: See sidebar for help.)

- Turn off and collapse the **Average Sea-Surface Temperature** layer.
- Turn on and expand the **Sea-Level Anomaly** layer.

An *anomaly* is anything that deviates from normal or average. In this layer, white represents places where sea level is at its normal height of 0 meters. Areas where sea level is higher than normal are colored pink and red, and areas that are lower than normal are colored blue. Next, you will see if the sea-level anomaly data support your prediction.

- Turn on the **Eastern Pacific** layer.
- Turn on the **Western Pacific** layer.

These layers define the regions you will use to compare the eastern and western equatorial Pacific Ocean. You will begin by using the **Sea-Level Anomaly** layer to look for differences in sea level between the two areas.

Next, you will calculate the sea-level anomaly for the western Pacific region.

- Click the Select By Location button.
- In the Select By Location window, construct the query statement:

  I want to **select features from** the **Sea-Level Anomaly** layer that **intersect** the features in the **Western Pacific** layer.

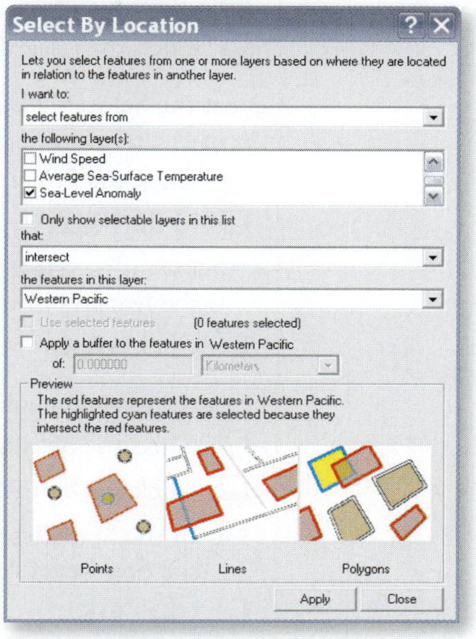

- Click **Apply.**
- Close the Select By Location window.

The western Pacific region will be highlighted.

- Click the Statistics button.
- In the Statistics window, calculate statistics for **only selected features** of the **Sea-Level Anomaly** layer, using the **Anomaly (cm)** field.

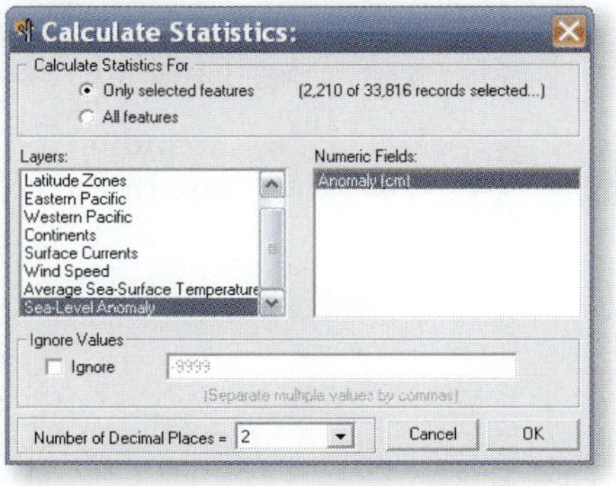

- Click **OK**.

The average sea-level anomaly for the western Pacific region is given in centimeters as the **Mean** in the Statistics window.

5. Round the average sea-level anomaly for the region to the nearest 0.1 cm and record it in Table 1 on the following page.

Table 1 — Sea-level anomalies in the equatorial Pacific Ocean

Western Pacific Anomaly *cm*	Eastern Pacific Anomaly *cm*	Difference (W Pacific Anomaly – E Pacific Anomaly) *cm*

- Close the Statistics window.
- Click the Clear Selected Features button.

Repeat the Select By Location procedure to calculate the sea-level anomaly for the eastern Pacific.

- Click the Select By Location button.
- In the Select By Location window, construct the query statement:

  I want to **select features from** the **Sea-Level Anomaly** layer that **intersect** the features in the **Eastern Pacific** layer.
- Click **Apply.**

The eastern Pacific region will be highlighted.

- Close the Select By Location window.
- Click the Statistics button.
- In the Statistics window, calculate statistics for **only selected features** of the **Sea-Level Anomaly** layer, using the **Anomaly (cm)** field.
- Click **OK**.

The average sea-level anomaly for the eastern Pacific region is given in centimeters as the **Mean** in the Statistics window.

6. Round the average sea-level anomaly for the region to the nearest 0.1 cm and record it in Table 1.
7. Calculate the difference between the average sea-level anomaly in the western Pacific and the eastern Pacific and record it in Table 1.

- Close the Statistics window.
- Click the Clear Selected Features button.

8. According to Table 1, is sea level higher in the western Pacific or in the eastern Pacific? Compare this result with your prediction in question 4.

This difference represents a low but very broad "hill" of water in the western half of the Pacific Ocean. This hill is essentially permanent, even though gravity is constantly pulling the hill down and outward.

Pacific Ocean profile

To understand the conditions in the Pacific Ocean, it may help to look at a cross section of the ocean and surface winds. Click the Media Viewer button and open the **Pacific Ocean Profile** image.

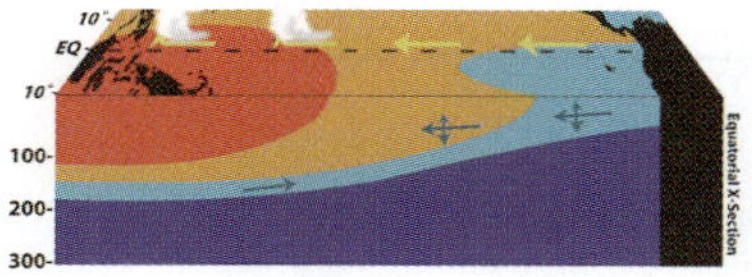

9. Explain how the prevailing winds and ocean currents might help keep the hill of water in the Pacific Ocean from spreading out. (See sidebar at left.)

The Southern Oscillation

Long before they understood the causes of El Niño, scientists noticed seemingly independent weather phenomena in the Southern Hemisphere. One of these was an oscillation, or periodic strengthening and weakening, of the surface atmospheric pressure across the Pacific Ocean. Scientists now understand that this oscillation and the sea-surface temperature changes characteristic of El Niño and La Niña phases are directly related. Together, these phenomena are now known as the *El Niño Southern Oscillation*, or ENSO.

In this section, you will compare the average atmospheric pressure in the equatorial eastern and western Pacific Ocean for three different oscillation phases: the "normal" long-term average calculated with data from *all* years between 1950 and 2003, eight years of El Niño, and eight years of La Niña.

- Click the QuickLoad button.
- Select **Data Frames**, choose **Southern Oscillation**, and click **OK**.

The **Normal Surface Atmospheric Pressure** layer represents the average surface pressure for all years from 1950 to 2003. Low pressure is represented by light shades of blue, and high pressure is shown in darker shades of blue. Surface pressure decreases as the air temperature and humidity increase.

- Click the QuickLoad button.
- Select **Spatial Bookmarks**, choose **Pacific Ocean**, and click **OK**.

Next, you will calculate the normal average surface pressure for the eastern Pacific region.

- Click the Select By Location button.
- In the Select By Location window, construct the query statement:

  I want to **select features from** the **Normal Surface Atmospheric Pressure** layer that **intersect** the features in the **Eastern Pacific** layer.
- Click **Apply**.

The eastern Pacific region will be highlighted.

- Close the Select By Location window.
- Click the Statistics button.
- In the Statistics window, calculate statistics for **only selected features** of the **Normal Surface Atmospheric Pressure** layer, using the **Pressure (mb)** field.
- Click **OK**.

What does "mb" stand for?

Millibar is abbreviated as *mb*. A bar (from the Greek *baros*, meaning *weight*) is a unit of pressure. The standard metric prefix for one thousandth is *milli*, so one millibar equals 0.001 bar. Standard atmospheric pressure is 1013.2 mb.

The average surface pressure for the normal phase in the eastern Pacific is given in millibars as the **Mean** in the Statistics window.

10. Round the average (**Mean**) surface pressure for the normal phase to the nearest 0.1 mb and record it in Table 2 on the following page. Data for the western Pacific have been entered for you.

Table 2 — Equatorial Pacific average surface pressure and climate phases

Region	Average surface atmospheric pressure *mb*		
	Normal phase	El Niño phase	La Niña phase
Western Pacific	1009.7	1010.1	1009.2
Eastern Pacific			
Difference (E – W)			

- Close the Statistics window.
- Click the Clear Selected Features button.
- Turn off and collapse the **Normal Surface Atmospheric Pressure** layer.
- Turn on and expand the **El Niño Surface Atmospheric Pressure** layer.

Repeat the Select By Location query process to calculate the average surface pressure for the eastern Pacific region during El Niño phases.

- Click the Select By Location button.
- In the Select By Location window, construct the query statement:

  I want to **select features from** the **El Niño Surface Atmospheric Pressure** layer that **intersect** the features in the **Eastern Pacific** layer. (Note: Make sure the **El Niño Surface Atmospheric Pressure** is the ONLY option checked.)
- Click **Apply**.

The eastern Pacific region will be highlighted.

- Close the Select By Location window.
- Click the Statistics button.
- In the Statistics window, calculate statistics for **only selected features** of the **El Niño Surface Atmospheric Pressure** layer, using the **Pressure (mb)** field.
- Click **OK**.

The average surface pressure during El Niño phases is given in millibars for the eastern Pacific as the **Mean** in the Statistics window.

11. Round the average El Niño surface pressure for the eastern Pacific to the nearest 0.1 mb and record it in Table 2.

- Close the Statistics window.
- Click the Clear Selected Features button.
- Turn off and collapse the **El Niño Surface Atmospheric Pressure** layer.
- Turn on and expand the **La Niña Surface Atmospheric Pressure** layer.
- Click the Select By Location button.
- In the Select By Location window, construct the query statement:

  I want to **select features from** the **La Niña Surface Atmospheric Pressure** layer that **intersect** the features in the **Eastern Pacific** layer. (Note: Make sure the **La Niña Surface Atmospheric Pressure** is the ONLY option checked.)

Click **Apply**.

The eastern Pacific region will be highlighted.

Close the Select By Location window.

Click the Statistics button.

In the Statistics window, calculate statistics for **only selected features** of the **La Niña Surface Atmospheric Pressure** layer, using the **Pressure (mb)** field.

Click **OK**.

The average surface pressure for the La Niña phase is given in millibars for the eastern Pacific as the **Mean** in the Statistics window.

12. Round the La Niña average surface pressure for the eastern Pacific to the nearest 0.1 mb and record it in Table 2.

Close the Statistics window.

Click the Clear Selected Features button.

13. Calculate the differences between the eastern and western Pacific regions for the three phases and record them in Table 2.

Winds blow from areas of high pressure to areas of low pressure, and the wind speed increases as the pressure differences increase. With this in mind, consider how the pressure difference between the western and eastern Pacific Ocean influences wind speeds during El Niño and La Niña phases.

14. Write the name *El Niño* or *La Niña* in each blank to correctly complete the following statement:

    The pressure difference between the eastern Pacific and the western Pacific is greatest during the __________ phase, resulting in higher wind speeds than in the normal and __________ phases.

Turn off and collapse the **La Niña Surface Atmospheric Pressure** layer.

Next, you will measure wind speeds in the equatorial Pacific Ocean to see whether wind speeds during El Niño and La Niña phases respond to the east-west pressure difference in the way you predicted.

Turn on and expand the **Wind Speed** layer.

Click the Select By Location button.

In the Select By Location window, construct the query statement:

I want to **select features from** the **Wind Speed** layer that **intersect** the features in the **Eastern Pacific** layer.

Click **Apply**, but do not close the Select By Location window.

The eastern Pacific region will be highlighted. This process selected the average wind speed data for the eastern Pacific. Next, you will add the data for the western Pacific so you can calculate the average wind speed across the entire equatorial Pacific.

In the Select By Location window, construct the query statement:

I want to **add to the currently selected features in** the **Wind Speed** layer that **intersect** the features in the **Western Pacific** layer.

Click **Apply**.

Both the eastern and western Pacific regions should be highlighted.

Close the Select By Location window.

Click the Statistics button.

In the Statistics window, calculate statistics for **only selected features** of the **Wind Speed** layer, using the **El Niño Mean Wind Speed**, **La Niña Mean Wind Speed**, and **Normal Mean Wind Speed** fields. (Note: Hold down the Shift key to select multiple fields.)

Click **OK**.

The average wind speeds (m/s) for the normal, El Niño, and La Niña phases are given for the entire equatorial Pacific region as the **Mean** in the Statistics window.

15. Round the average (**Mean**) wind speeds to the nearest 0.01 m/s and record them in Table 3.

Table 3 — Equatorial Pacific Ocean wind speed comparison

Average wind speed *m/s*		
Normal phase	El Niño phase	La Niña phase

Close the Statistics window.

Click the Clear Selected Features button.

16. Do the average wind speeds you recorded in Table 3 agree with the predictions you made based on pressure in question 14 about the wind speeds in La Niña years? Explain.

The effects of El Niño and La Niña on sea level

Now that you have seen how surface atmospheric pressure and wind speed vary during El Niño and La Niña phases, think back to the concept of sea level you explored earlier. As you observed, the warm waters of the equatorial western Pacific form a hill or mound due to thermal expansion, and are held in place by the trade winds and equatorial currents. You have also seen that trade winds strengthen or weaken in response to changes in surface pressure and ocean temperature associated with El Niño and La Niña phases.

Next you will examine how the strengthening or weakening of trade winds affects sea level.

Turn off the **Wind Speed** layer.

Turn on the **Sea-Level Stations** layer.

This layer shows the locations of sea-level monitoring stations in the Pacific Ocean. The points in the middle of the Pacific are located on islands.

Click the QuickLoad button.

Select **Layers**, choose **El Niño Sea-Level Anomalies**, and click **OK**.

A new **Sea-Level Stations** layer will appear at the top of the Table of Contents with the subheading, **Ave El Niño MSL anomaly**.

Turn on the **Sea-Level Stations** layer with the subheading **Ave El Niño MSL anomaly.**

The sea-level stations in this new layer are classified based on whether sea level during El Niño phases is below normal (blue) or above normal (red). In this case, normal is the mean sea level at that location.

17. Based on what you have learned about the effect of El Niño on trade winds and sea level, complete the following statement by circling the appropriate words where indicated and writing *higher* or *lower* in each blank.

    During an El Niño phase, the speed of the trade winds is _________ than in the normal phase, causing the hill of warm water in the western Pacific to pile up / spread out (circle one). As a result, sea level is _________ than normal in the western Pacific and ________ than normal in the eastern Pacific.

Turn off the **Sea-Level Stations** layer with the subheading **Ave El Niño MSL anomaly.**

Click the QuickLoad button.

Select **Layers**, choose **La Niña Sea-Level Anomalies**, and click **OK**.

A new **Sea-Level Stations** layer will appear at the top of the Table of Contents with the subheading, **Ave La Niña MSL anomaly**.

Turn on the **Sea-Level Stations** layer with the subheading **Ave La Niña MSL anomaly.**

The sea-level stations in this layer are classified based on whether sea level during La Niña phases is below normal (blue) or above normal (red). In this case, normal is the mean sea level at that location.

18. Based on what you have learned about the effect of La Niña on trade winds and sea level, complete the following statement by circling the appropriate words where indicated and writing *higher* or *lower* in each blank.

    During La Niña phases, the speed of the trade winds is _________ than in the normal phase, causing the hill of warm water in the western Pacific to pile up / spread out (circle one). As a result, sea level is _________ than normal in the western Pacific and _________than normal in the eastern Pacific.

Turn off and collapse all **Sea-Level Stations** layers.

Effects of El Niño and La Niña on sea-surface temperature

Now you will explore how El Niño and La Niña affect sea-surface temperatures.

Turn on and expand the **El Niño SST Anomaly** layer.

This layer shows sea-surface temperature anomalies averaged over eight years of the El Niño phase. Reds represent positive anomalies, where the temperature was higher than normal; and blues represent negative anomalies, where the temperature was lower than normal.

19. In the **El Niño SST Anomaly** layer, where are large, positive sea-surface temperature anomalies (warming) more common — in the eastern or the western Pacific Ocean?

- Turn off and collapse the **El Niño SST Anomaly** layer.
- Turn on and expand the **La Niña SST Anomaly** layer.

20. In the **La Niña SST Anomaly** layer, where are large, negative temperature anomalies (cooling) more common — in the eastern or in the western Pacific Ocean?

Now you will quantify the sea-surface temperature anomalies across the equatorial Pacific Ocean.

- Click the Select By Location button.
- In the Select By Location window, construct the query statement:

  I want to **select features from** the **El Niño SST Anomaly** layer that **intersect** the features in the **Western Pacific** layer.
- Click **Apply.**
- Close the Select By Location window.
- Click the Statistics button.
- In the Statistics window, calculate statistics for **only selected features** of the **El Niño SST Anomaly** layer, using the **SST Anomaly (C)** field.
- Click **OK**.

The average sea-surface temperature anomaly for the El Niño phase is given in degrees Celsius for the western Pacific region as the **Mean** in the Statistics window.

21. Record the average (**Mean**) SST anomaly for the western Pacific region in the El Niño phase in Table 4. Round to the nearest 0.01 °C.

Table 4 — SST anomalies for El Niño and La Niña phases

Phase	SST Anomaly °C	
	Western Pacific	Eastern Pacific
El Niño		
La Niña		

- Close the Statistics window.
- Click the Clear Selected Features button.
- Repeat the Select By Location and Statistics processes to determine the SST anomaly for the **El Niño SST Anomaly** layer in the **Eastern Pacific** and record your answer in Table 4.

- Repeat the Select By Location and Statistics processes to determine the SST anomaly for the **La Niña SST Anomaly** layer in the **Western Pacific** and record your answer in Table 4. (Note: Make sure the **La Niña SST Anomaly** is the ONLY option checked.)
- Repeat the Select By Location and Statistics processes to determine the SST anomaly for the **La Niña SST Anomaly** layer in the **Eastern Pacific** to complete Table 4.
- When you are finished, close the Statistics and Select By Location windows.

22. Warming of the ocean surface produces increased evaporation and precipitation, whereas cooling decreases evaporation and precipitation. Answer the following questions on the basis of the data recorded in Table 4:

    a. Which part of the Pacific Ocean should experience the largest increase in precipitation and evaporation? During which phase of the oscillation does this occur (El Niño or La Niña)?

    b. Which part of the Pacific Ocean should experience the largest decrease in precipitation and evaporation? During which phase of the oscillation does this occur? (Normal, El Niño, or La Niña)

- Turn off and collapse the **La Niña SST Anomaly** layer.
- Turn off the **Eastern Pacific** and **Western Pacific** layers.
- Click the Clear Selected Features button.

Weather and climate effects

In addition to producing predictable weather and climate patterns in the equatorial Pacific, ENSO also influences weather and climate around the world. To illustrate this, you will examine the effects of El Niño and La Niña on global precipitation rates and drought patterns.

- Click the Zoom to Full Extent button to view the entire map.
- Turn on and expand the **El Niño Precipitation Anomaly** layer.

This layer shows the deviation from normal precipitation averages during the El Niño phase, in centimeters per year. Negative anomalies (below average values) are shown in orange, and positive anomalies (above average values) are shown in purple.

- Turn off and collapse the **El Niño Precipitation Anomaly** layer.
- Turn on and expand the **La Niña Precipitation Anomaly** layer.

This layer shows the deviation from normal precipitation averages during La Niña phase, in centimeters per year. Again, negative anomalies (below average values) are shown in orange, and positive anomalies (above average values) are shown in purple.

What does "PDSI" stand for?

PDSI is a numerical value calculated using several climate measurements, but has no units itself. It is useful for comparing conditions of different regions or of the same region at different times.

23. Having viewed the maps of precipitation anomalies, what can you say about global precipitation patterns during El Niño and La Niña phases?

- Turn off and collapse the **La Niña Precipitation Anomaly** layer.

ENSO and climate

To conclude your investigation, you will examine two layers that indicate whether the worldwide climate conditions were unusually dry or wet during El Niño or La Niña phases.

- Turn off the **Continents** layer.
- Turn on and expand the **El Niño PDSI** layer.

The Palmer Drought Severity Index (PDSI) is a measure of the relative wetness or dryness of an area. This layer shows PDSI values for areas around the world. This is also an anomaly layer—negative values, in brown, represent drought conditions; and positive values, in green, represent wetter than normal conditions. Next, you will identify countries that are wetter or drier than average during El Niño phases.

- To determine the name of a country, click the Identify tool.
- In the Identify Results window, select the **Countries** layer from the drop-down list of layers.
- Next, click within the borders of the country you wish to identify.

24. Identify three countries that appear to be wetter than average during El Niño phases, and three that appear to be drier than average.
    a. Wetter-than-average countries.
        (1)
        (2)
        (3)
    b. Drier-than-average countries.
        (1)
        (2)
        (3)

Next, you will find the average PDSI values for six selected countries, to assess how each is affected during an El Niño phase.

- Close the Identify Results window.
- Click the Select By Attributes button.
- Choose **El Niño PDSI** from the Layers menu.
- Click **Quickload Query** in the Select By Attributes window, choose **El Niño PDSI by Country**. This will automatically enter the following query:

("NAME" = 'Argentina') OR ("NAME" = 'Australia') OR ("NAME"='Finland') OR ("NAME" = 'India') OR ("NAME" = 'South Africa') OR ("NAME" = 'United States')

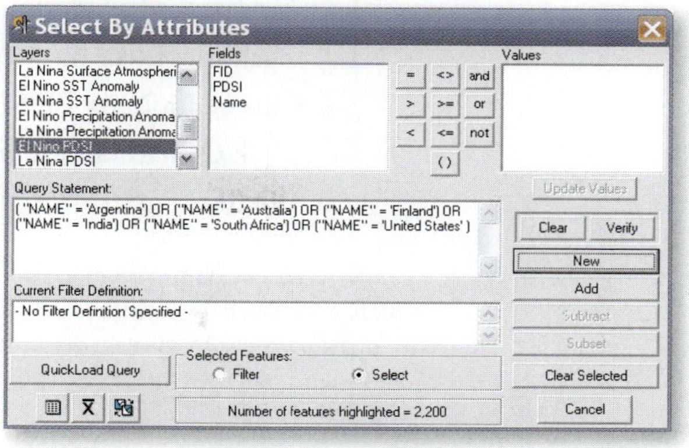

Click **New**.

The six selected countries should be highlighted. Now you will average the data for each country by creating a summary table.

Close the Select By Attributes window.

Click the Summarize button [Σ].

In the Summarize window, select **El Niño PDSI** as the **feature layer**.

Select **Name** as the **field to summarize** in the drop-down menu.

Double-click **PDSI** to display the statistics options and check **Average**.

Check the **Summarize on Selected Records only** option and click **OK**.

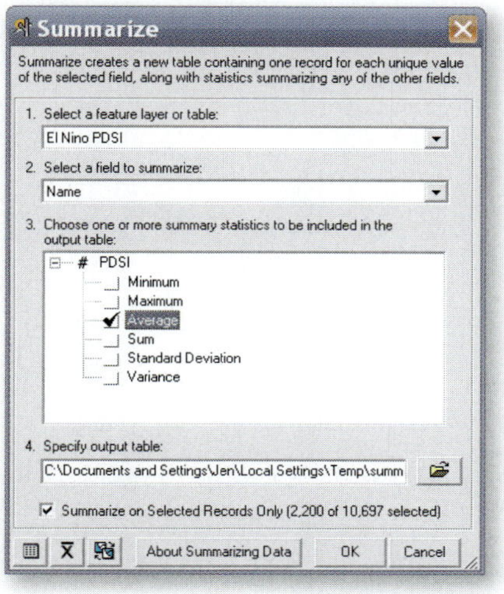

25. Round the average El Niño PDSI values for the six countries to the nearest 0.01 and record them in Table 5.

Table 5 — El Niño and La Niña PDSI values for selected countries

Country Name	PDSI value		
	El Niño	La Niña	Change +/–
Argentina			
Australia			
Finland			
India			
South Africa			
United States			

Close the Summary Table window.

Click the Clear Selected Features button.

Turn off and collapse the **El Niño PDSI** layer.

- Turn on and expand the **La Niña PDSI** layer.
- Click the Select By Attributes button.
- Choose **La Niña PDSI** from the Layers menu.
- Click **Quickload Query** in the Select By Attributes window, choose **La Niña PDSI by Country**. This will automatically enter the following query:

("NAME" = 'Argentina') OR ("NAME" = 'Australia') OR ("NAME"='Finland') OR ("NAME" = 'India') OR ("NAME" = 'South Africa') OR ("NAME" = 'United States')

- Click **New**.

The six selected countries should be highlighted. Now you will average the data for each country by creating a summary table.

- Close the Select By Attributes window.
- Click the Summarize button **Σ**.
- In the Summarize window, select **La Niña PDSI** as the **feature layer**.
- Select **Name** as the **field to summarize** in the drop-down menu.
- Double-click **PDSI** to display the statistics options and check **Average**.
- Check the **Summarize on Selected Records only** option and click **OK**.

26. Round the average La Niña PDSI results to the nearest 0.01 and record them in Table 5.
27. Write a "+" or "–" sign in the fourth column of Table 5 to indicate if the PDSI increases or deceases from the El Niño to La Niña phases in each country.
28. According to your results, which of the six countries are most likely to experience
    a. drought during El Niño years?

    b. wet La Niña years?

29. Summarize how the general patterns in drought and wet years vary with El Niño and La Niña.

- Close the Summary Table window.
- Click the Clear Selected Features button.
- Quit ArcMap and do not save changes.

Wrap-up 3.5

Local interactions

Now that you have examined some of the global effects during El Niño and La Niña phases, you will look at how these events affect the temperature and precipitation across the United States in general, and at a specific location of your choice.

- Launch ArcMap, locate and open the **etoe_unit_3.mxd** file.

Refer to the tear-out Quick Reference Sheet located in the Introduction to this module for GIS definitions and instructions on how to perform tasks.

- In the Table of Contents, right-click the **Regional Effects** data frame and choose Activate.
- Expand the **Regional Effects** data frame.

The **U.S. Climate** layer contains average precipitation and temperature data for different regions of the United States during normal, El Niño, and La Niña phases. In order to visualize these data, you will load new layers each showing a different climate measurement.

- Click the QuickLoad button.
- Select **Layers**, click **El Niño Precipitation Anomaly**, and click **OK**.
- Repeat the Quickload process to load the **El Niño Temperature Anomaly**, **La Niña Precipitation Anomaly**, and **La Niña Temperature Anomaly** layers.
- Turn off all of the newly loaded layers. Turn them on, one at a time, as needed to answer the question below.

1. For each layer, describe the pattern shown across the continental United States. For precipitation, describe states or regions that are wetter or drier than average. For temperature, describe states or regions that are warmer or cooler than average.

   a. El Niño Precipitation Anomaly

   b. El Niño Temperature Anomaly

   c. La Niña Precipitation Anomaly

   d. La Niña Temperature Anomaly

- Turn off the **El Niño Temperature Anomaly**, **La Niña Precipitation Anomaly**, and **La Niña Temperature Anomaly** layers.

- Choose any location in the United States on which you want to gather further climate data, and use the Zoom In tool to zoom in on it.
- Click the Identify tool.
- In the Identify Results window, select the **U.S. Climate By Region** layer from the drop-down list of layers.
- Next, click on the location you wish to identify.

2. Write the name of your chosen location in Table 1, then use the information in the Identify Results window to complete the table. Round the temperature and precipitation anomaly values to the nearest 0.01 unit.

Table 1 — Climate data for your chosen location

Information	Value
Name of region	
State	
(Normal) Avg Annual Precip (cm/yr)	
(Normal) Avg Annual Temp (°C)	
El Niño Precip Anomaly (cm/yr)	
El Niño Temp Anomaly (°C)	
La Niña Precip Anomaly (cm/yr)	
La Niña Temp Anomaly (°C)	

3. Using the average El Niño, La Niña, and Normal values from Table 1, describe how each of the following phases affects the climate (temperature, precipitation, etc.) of your chosen location.

   a. El Niño

   b. La Niña

- Close the Identify Results window.

Present conditions

Now that you have learned about the effects that El Niño and La Niña have on your chosen area, you will see what the present conditions are in the South Pacific Ocean.

Open the NOAA El Niño Web page by selecting NOAA El Niño using the Media Viewer button or by entering its address directly in your Web browser:

http://www.elnino.noaa.gov

To open the Web page when the **etoe_unit_3.mxd** project file is open:

- Click the Media Viewer button.
- Load the **NOAA El Niño** Web page.

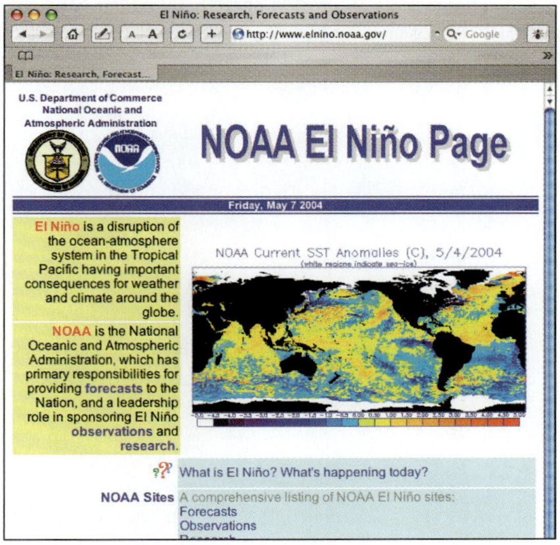

Broken links?

If the Web links in this activity do not work, alternate links may be found at:

http://www.scieds.com/saguaro/webresources.php?id=etoe

The NOAA El Niño Web page displays a map of the present sea-surface temperature anomalies. Examine the temperature anomalies as they exist today, and consider what you have learned about El Niño and La Niña phases.

4. Describe the pattern of temperature anomalies. Is it unusually warm in the east or west part of the equatorial Pacific Ocean, or are the temperatures as expected for normal-phase years?

El Niño forecasts

http://www.elnino.noaa.gov/forecast.html

Click the **Forecasts** link, then click the El Niño **status** link on the El Niño Forecasts page.

This takes you to the El Niño Southern Oscillation (ENSO) Diagnostic Discussion page containing the ongoing discussions among experts in the field of meteorology who study ENSO. Their language can be technical as they discuss different observations. Study the report and see if you can determine whether the experts are predicting the development of El Niño, La Niña, or normal phases. Use this information and the information on the main El Niño Web page to answer the following questions.

5. Do present sea-surface temperature anomalies, as shown on the NOAA Web site, suggest an El Niño phase, a La Niña phase, or normal phase conditions?

Normal or neutral?

Some sources refer to the normal phase as the *neutral* phase. The two terms may be used interchangeably.

6. Using these assessments by experts, would you predict that a normal, an El Niño, or a La Niña phase might develop in the near future?

7. On the basis of those expert assessments, determine what the near future may hold in terms of precipitation and temperature for your chosen area.

- Close your Web browser.
- Quit ArcMap and do not save changes.

Unit 4
Marine Productivity

In this unit, you will

- *Discover patterns in global primary productivity.*
- *Compare terrestrial and marine productivity.*
- *Explore the key resources required for productivity.*
- *Correlate variations in marine productivity with limiting resources.*
- *Investigate sources of marine nutrients.*
- *Synthesize observations to evaluate the causes of dead zones.*

NOAA

The ocean provides up to 20 percent of the world's food supply.

Warm-up 4.1

Bounty from the sea

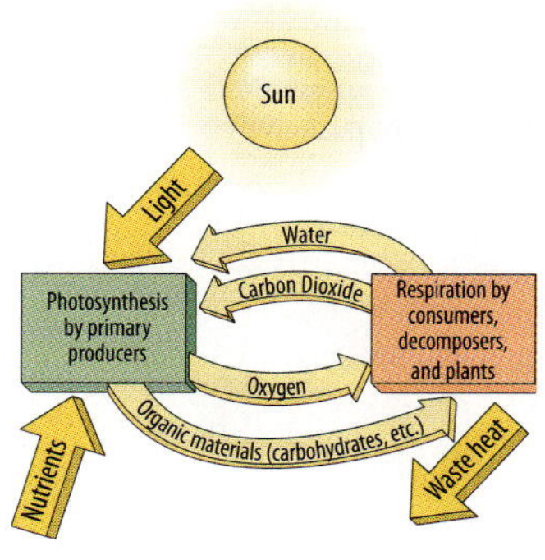

Figure 1. Photosynthesis and respiration.

autotroph — ("self-feeder") organism that makes its own food rather than consuming other organisms.

heterotroph — ("other-feeder") organism that consumes other organisms.

Seafood makes up 20 percent of the world's food supply, with over one billion people depending on its resources for survival. As seafood harvests have increased over the past two centuries, populations of some species of marine life have decreased and even become extinct. Given the ocean's vast area, it is difficult to locate, monitor, and track changes in stocks of commercially important fish and shellfish. Thus, scientists frequently use satellites to indirectly assess the health of fisheries and the ocean ecosystem.

A key indicator of the ocean's health is *primary productivity*, or the rate at which new organic material is produced through *photosynthesis* (Figure 1 at left). Photosynthesis is the process by which plant cells containing the green pigment *chlorophyll* use sunlight to convert water and carbon dioxide into the food (sugars and starches) and oxygen needed by most other organisms. Satellites can measure the amount of chlorophyll contained in single-celled plants in the ocean's surface layer, from which we can estimate primary productivity.

Food chains (Figure 2 below) and more complex food webs (Figure 3 below) illustrate feeding relationships among organisms in biological communities. At the base of food chains and webs are *autotrophs*, which produce their own food for growth and reproduction through photosynthesis. In the ocean, the primary autotrophs are phytoplankton, microscopic single-celled plants that drift near the ocean surface. The remaining organisms in food webs are *heterotrophs*, which obtain food by feeding on other organisms.

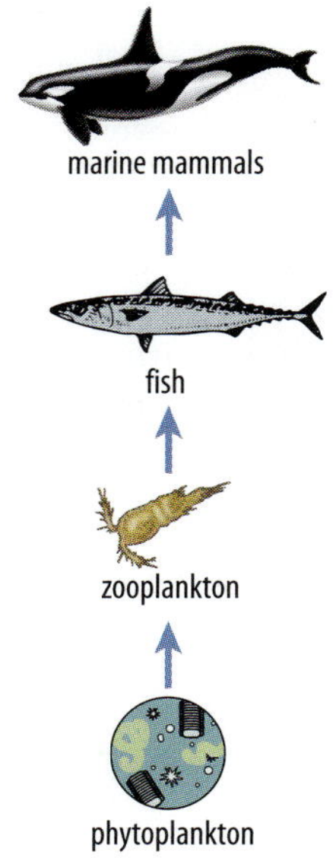

Figure 2. Simple marine food chain. Arrows represent the transfer of energy from one organism to another through consumption.

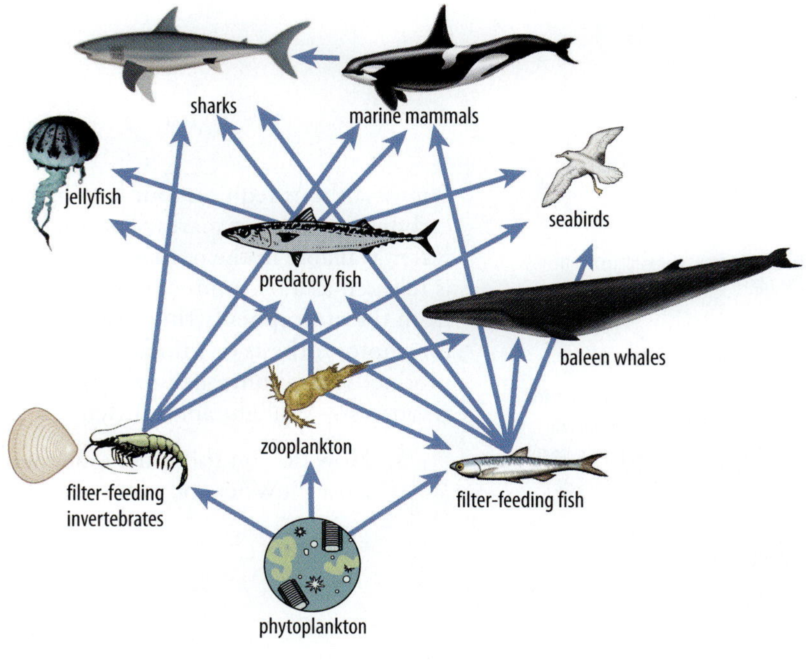

Figure 3. A marine food web.

biotic community — group of interdependent organisms inhabiting the same region and interacting with one another.

anchovies — a type of small fish, similar to sardines, that eat zooplankton.

tuna — a type of large predatory fish that eat other fish.

The preservation of each link in a food web is critical for maintaining diverse and healthy *biotic communities*. However, certain organisms are more critical than others.

1. Examine the complex marine food web in Figure 3 on the previous page. Add anchovies (**A**), tuna (**T**), and humans (**H**) where you think they fit best. Draw arrows as needed to show consumption of and by other organisms.

2. What do you think would happen if all of the autotrophs were removed from the marine food web?

3. What do you think would happen if one of the heterotrophs, such as the predatory fish, were removed?

4. How do humans influence food webs?

respiration — process by which organisms oxidize or "burn" food, producing water and carbon dioxide.

nutrients — chemical compounds that are used by bacteria and plants as the building blocks for organic material. Common nutrients include:

- phosphates (PO_4^-)
- nitrates (NO_3^-)
- silica (SiO_4^-)
- iron (Fe^{3+})

Photosynthesis requires four key ingredients: water, sunlight, nutrients, and carbon dioxide. For marine autotrophs like phytoplankton, there is plenty of water available in the oceans. Carbon dioxide is also abundant in ocean waters. It is released as a by-product of *respiration*, and it is readily absorbed into the ocean from the atmosphere. Thus, water and carbon dioxide are not limiting resources for photosynthesis or primary productivity in the ocean. However, marine productivity is controlled or limited by the availability of the other two necessary resources — sunlight and nutrients.

5. How do you think the availability of limiting resources might vary in the ocean? (Where or when would they be high or low?)

6. Considering the availability of limiting resources, where would you expect phytoplankton to be most productive or least productive? (Near the equator or near the poles? Near the coast or in the open ocean?)

7. How is the productivity of autotrophs important to society? Describe three different ways in which a severe decrease in primary productivity would affect society.

   a.

   b.

   c.

Investigation 4.2

The life-giving ocean

Photosynthesis

The general chemical equation for photosynthesis is:

$$6\,CO_2 + 6\,H_2O \xrightarrow{\text{sunlight}} C_6H_{12}O_6 + 6\,O_2$$

carbon dioxide — water — glucose (carbohydrate) — oxygen

NASA

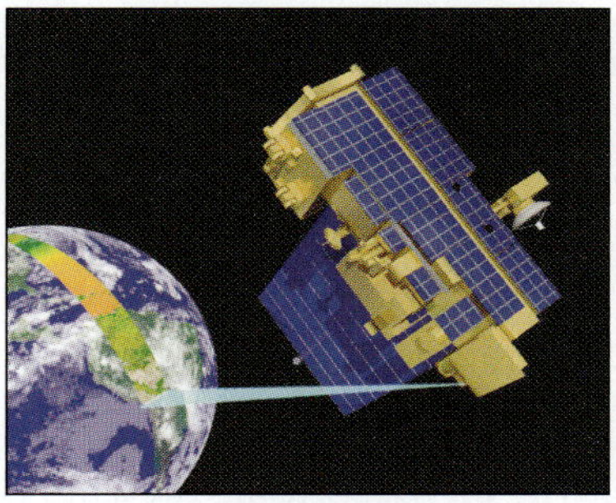

Figure 1. The MODIS (MODerate-resolution Imaging Spectrometer) instrument on the Terra satellite is the latest tool for measuring primary productivity from space.

primary productivity — the rate at which new organic material is formed by photosynthesis.

Phytoplankton are tiny—several of them could fit side-by-side across the width of a human hair—but collectively, they pack a wallop. Phytoplankton are *primary producers*, serving as the first link in almost every food chain in the ocean. They transform water and carbon dioxide into carbohydrates, which they use for producing energy for growth and reproduction. Phytoplankton, in turn, are food for other organisms, passing carbohydrates and other nutrients up the food chain. Because phytoplankton release oxygen during photosynthesis, they also play a significant role in maintaining the proper balance of Earth's atmospheric gases. Phytoplankton produce about half of the world's oxygen and, in doing so, remove large amounts of carbon dioxide from the atmosphere.

Given the central role of phytoplankton in stabilizing the mixture of gases in Earth's atmosphere and in providing food for other organisms, it is important to monitor their location and rate of productivity. Although it is impossible to directly measure the productivity of phytoplankton on a global scale, there are several ways of making indirect calculations. One method relies on the distinctive way that chlorophyll, the green pigment in phytoplankton and other plants, reflects sunlight. By using satellites to measure the chlorophyll concentration of the ocean surface layer, scientists can estimate the rate at which phytoplankton produce carbohydrates (Figure 1). Because carbon is the key element in the process, productivity is measured in terms of kilograms of carbon converted per square meter of ocean surface (kgC/m^2) per year.

Global primary productivity

In this exercise you will examine global *primary productivity* and its relation to the factors that support phytoplankton growth.

- Launch ArcMap, and locate and open the **etoe_unit_4.mxd** file.

Refer to the tear-out Quick Reference Sheet located in the Introduction to this module for GIS definitions and instructions on how to perform tasks.

- In the Table of Contents, right-click the **Primary Productivity** data frame and choose Activate.
- Expand the **Primary Productivity** data frame.

This data frame shows the average annual primary productivity for terrestrial and marine environments. During their respective winters, regions near the north and south poles receive little or no sunlight, making satellite measurements impossible. Nonetheless it is reasonable to assume that, with no sunlight, winter primary productivity in these regions is essentially zero.

1. What colors represent areas of highest and lowest productivity?

   a. Highest

   b. Lowest

2. On Map 1, circle the land areas with the highest productivity using solid lines, and the land areas with the lowest productivity using dashed lines.

Map 1 — Areas of highest and lowest productivity

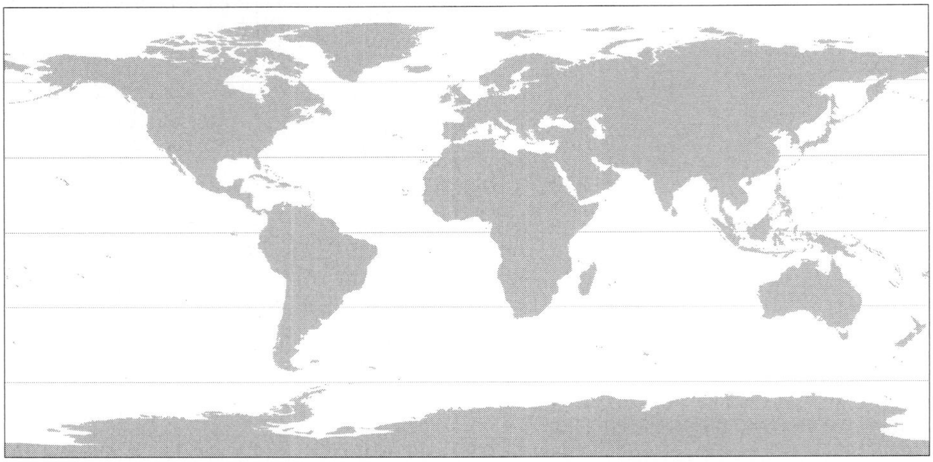

- Turn off the **Terrestrial Productivity** layer.
- Using the Zoom In tool, examine, in detail, the areas with the highest marine productivity.
- When you are finished, click the Full Extent button to zoom back out to show the entire map.

3. Mark the ocean areas with highest productivity on Map 1 using the label **H** (high), and the ocean areas of lowest productivity using the label **L** (low).

4. Where is marine productivity generally

   a. highest?

   b. lowest?

5. Compare the locations of regions of high terrestrial and marine productivity on Map 1. Describe any geographic patterns or similarities in their distribution.

Productivity and distance from coast

Next, you will examine productivity near the coastline in greater detail.

- Turn off the **Marine Productivity** layer.

- Turn on and expand the **Coastal Productivity** layer.
- Click the Summarize button **Σ**.
- In the Summarize window, select **Coastal Productivity** as the **feature layer.**
- Select **Distance** as the **field to summarize** in the drop-down menu.
- Double-click **Area (m^2)** to display the statistics options, check **Sum.**
- Next, double-click **Total Productivity (kgC)** to display the statistics options, check **Sum,** and click **OK**. (Be patient—summary tables may take a while to process.)

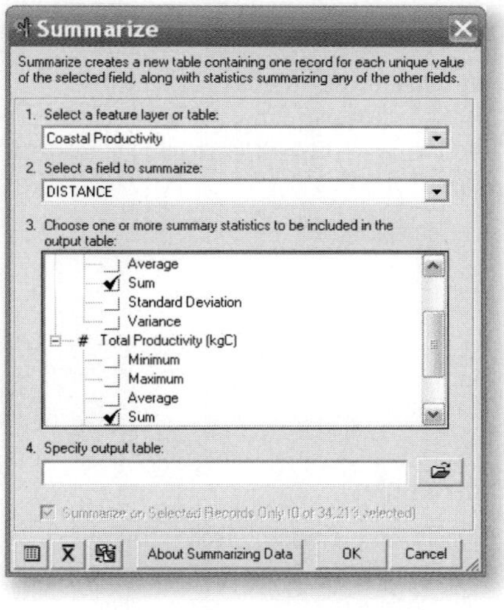

In the resulting summary table, the **Sum TOTAL_Prod** field gives the total productivity, and the **Sum_PROJECTED** field gives the total area, for each coastal zone. These values are very large, in the trillions. One trillion is the word for the number value 1,000,000,000,000.

Converting to trillions

Move the decimal point 12 places to the left, then round to the appropriate decimal place.

Example:

27656585301787.3

= 27.7 trillion

6. Record the total productivity and total area for each coastal zone in Table 1. Round productivity and area to the nearest 0.1 trillion. (See sidebar for help with converting values to trillions.)

Table 1 — Marine productivity with distance from coastline

Coastal zone	Total productivity *trillion kgC*	Total area *trillion m^2*	Average productivity *kgC/m^2*
Open Ocean (> 960 km)	27.7	211.0	0.13
Far (640 – 960 km)			
Mid (320 – 640 km)			
Near (0 – 320 km)			

Calculating average productivity

To calculate average productivity, divide the Total Productivity by the Total Area. Be sure to write both measurements in trillions, so the trillions will cancel each other.

Example:

$$\frac{\text{Total Productivity}}{\text{Total Area}} = \frac{27.7\ \text{~~trillion~~ kgC}}{211.0\ \text{~~trillion~~ m}^2}$$

$= 0.13\ kgC/m^2$

7. Calculate the average marine productivity for each zone and record your results in Table 1. Round to the nearest 0.01 kgC/m^2. (See sidebar for help calculating average productivity.)

8. What happens to the level of marine productivity as the distance seaward from the coastline increases?

9. Of the resources necessary for photosynthesis (water, sunlight, carbon dioxide, and nutrients), which do you think is most likely to change with distance from the coast to produce the pattern you observe in Table 1? Explain your answer.

- Close the Summary Table window.

Understanding the patterns

In this section, you will determine when and where sunlight and nutrients control marine productivity. You will examine nitrates and phosphates, but other *nutrients* like silica and iron affect productivity in similar ways.

nutrients — chemical compounds that are used by bacteria and plants as the building blocks for organic material. Common nutrients include:

- phosphates (PO_4^-)
- nitrates (NO_3^-)
- silica (SiO_4^-)
- iron (Fe^{3+})

- Turn off and collapse the **Coastal Productivity** layer.
- Turn on the **Marine Nutrients** layer group.
- Turn on and expand the **Nitrates** layer.

Nitrates and phosphates are important nutrients that are used by autotrophs for building complex molecules needed for growth and development. The **Nitrates** layer displays the average annual level of nitrates in the world's oceans in terms of *micromolarity*, or millionths of a mole of nitrate per liter of seawater (see sidebar).

Micromoles

Micromolarity is a measure of concentration used to describe very weak solutions. The metric prefix *micro* represents one millionth. A *mole* is 6.02×10^{23} molecules (or atoms), so a micromole is one millionth of a mole, or 6.02×10^{17} molecules. That seems like a lot, but when dissolved in a liter of water (55.5 moles, or about 3.34×10^{25} water molecules) it's only one molecule of nutrient for every 200 million water molecules. That's a weak solution.

10. Which latitude bands have the highest and lowest concentrations of nitrates? (See *Latitude bands* sidebar for help.)

    a. Highest.

    b. Lowest.

Latitude bands

Low latitudes

0° – 30° N and S
near equator

Middle latitudes

30° – 60° N and S
between equator and poles

High latitudes

60° – 90° N and S
near poles

- Turn off and collapse the **Nitrates** layer.
- Turn on and expand the **Phosphates** layer.

The **Phosphates** layer shows the average annual concentration of phosphates in the world's oceans in terms of micromolarity.

11. Which latitude bands have the highest and lowest concentration of phosphates?

    a. Highest.

    b. Lowest.

12. Are the patterns for both nutrients similar? If not, how do they differ?

- Turn off and collapse the **Phosphates** layer.
- Turn on the **Solar Radiation Flux** layer.

This layer shows the average annual solar radiation that strikes Earth's surface, in watts per square meter (W/m^2).

- Click the Media Viewer button and open the **Solar Flux Movie**.

This animation shows changes in solar radiation throughout the year. The time of year and the legend appear at the bottom of the image.

View the movie several times.

Use the **Solar Radiation Flux** layer and the **Solar Flux Movie** to answer the following questions.

13. Near what latitude is the average solar radiation throughout the entire year
    a. highest?
    b. lowest?

14. How does the pattern of nutrient concentration you noted in questions 10 and 11 compare to the pattern of solar radiation? Explain your answer.

Close the Media Viewer window.

Click the Media Viewer button, and open the **Productivity Movie**.

This animation shows marine productivity throughout the year. The time of year is indicated at the top of the image, and the legend appears at the bottom. Black areas at high latitudes are where there was no sunlight during the winter; you may assume these areas have zero productivity.

Study the movie to examine how productivity changes throughout the year. Focus on only the extreme high and low levels of productivity. You will need to play the movie several times to fill in Table 2.

15. In Table 2, enter the months and season when productivity is highest and lowest in each hemisphere. (See sidebar for help with the seasons.)

Table 2 — Productivity extremes by hemisphere

Hemisphere	Months of high productivity	Season	Months of low productivity	Season
Northern				
Southern				

Close the Media Viewer window.

There is very little seasonal variability in nutrients in the high-latitude oceans because the *thermocline* or temperature gradient between shallow and deep water is small, due to low solar-radiation input. This allows free circulation between deep nutrient-rich water and shallow nutrient-poor water. The year-round strong solar radiation in equatorial regions creates a strong thermocline (warm shallow water and cold deep water), which inhibits shallow- and deep-water circulation. Equatorial upwelling does bring nutrients to the surface in some areas, but

Earth's Seasons are

- caused by the tilt of Earth's axis.
- opposite in the northern and southern hemispheres.

Dates (typical)	Hemisphere	
	Northern	Southern
Dec 21 – Mar 20	Winter	Summer
Mar 20 – Jun 21	Spring	Fall
Jun 21 – Sep 22	Summer	Winter
Sep 22 – Dec 21	Fall	Spring

nutrient levels are generally low at low latitudes, particularly during the summer when the air is very stable and winds are weak.

16. Discuss the influence that each of the following factors has on productivity at *high* latitudes, and how productivity in high-latitude regions may change with the seasons.

    a. Sunlight

    b. Nutrients

17. Discuss how sunlight and nutrient levels contribute to the productivity pattern at *low* latitudes, and how productivity in that region may change with the seasons.

    a. Sunlight

    b. Nutrients

In the next activity, you will learn more about marine productivity, the sources of marine nutrients, and the processes that bring nutrients to the surface near the coastlines and in the open ocean.

Quit ArcMap and do not save changes.

Reading 4.3

Resources for productivity

Mean, green food-making machines

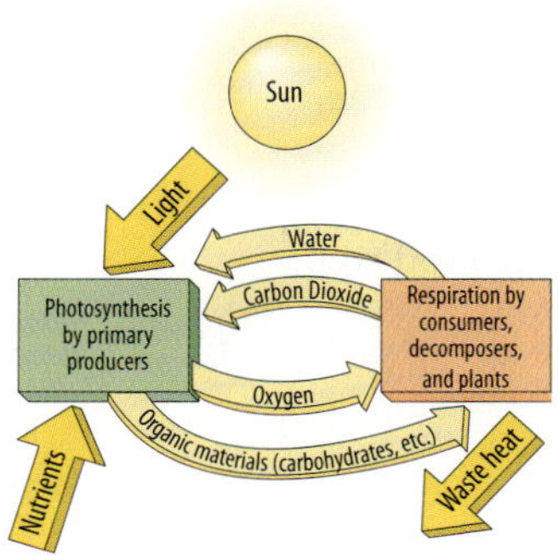

Figure 1. Photosynthesis and respiration.

With few exceptions, life on Earth depends on **photosynthesis**, the biological process that converts solar energy and inorganic compounds into food. Only autotrophs or "producers," including green plants and phytoplankton, are capable of photosynthesis. They contain the pigment **chlorophyll**, which uses solar energy to convert carbon dioxide and water into carbohydrates (Figure 1 at left). The amount of carbon converted to food by autotrophs is referred to as **primary productivity**. Autotrophs use some of this food immediately, and store the remainder for later use, to be converted back to energy through the process of **respiration**. Phytoplankton are consumed by other organisms which are, in turn, consumed by other organisms up the food chain. Collectively, these consumer organisms are called **heterotrophs**.

NOAA

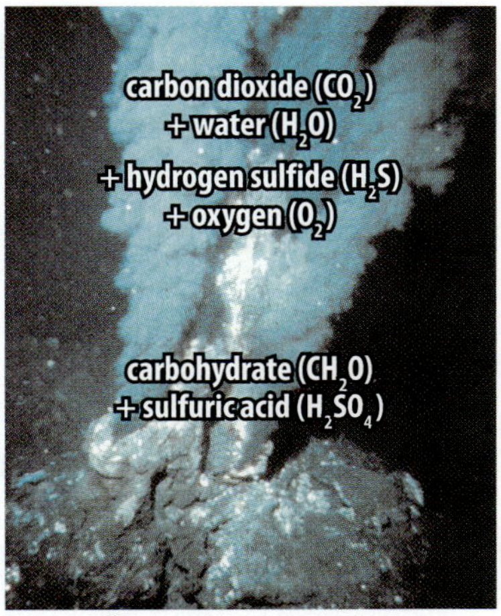

Figure 2. Process of chemosynthesis.

Exceptions to the rule

Green plants and phytoplankton are not the only organisms capable of synthesizing their own food. In the last 40 years, scientists have discovered biological communities that are not based on the sun's energy. The autotrophs in these communities are microbes that convert carbon dioxide and water into food using chemicals rather than sunlight. This process is called **chemosynthesis** (Figure 2 at left). Chemosynthetic bacteria thrive in the high temperatures and pressures of environments like deep-sea volcanic vents. They synthesize food using the chemical energy of sulfur compounds emerging from the vents. Chemosynthesis supports a diverse community of organisms (Figure 3 bottom left).

NOAA

Figure 3. Tube worms at hydrothermal vents consume chemosynthetic bacteria.

Food for thought—bottoms up!

Food webs illustrate the feeding relationships among organisms in a biotic community. The arrows represent the transfer of energy from one organism to another through consumption (Figure 4 at right). Autotrophs produce most of Earth's atmospheric oxygen as a by-product of photosynthesis, and are the foundation of nearly all terrestrial and aquatic food chains. Thus, primary productivity in terrestrial and marine environments is a key indicator of the overall health of the environment. Our ability to monitor primary productivity has profound economic and environmental importance.

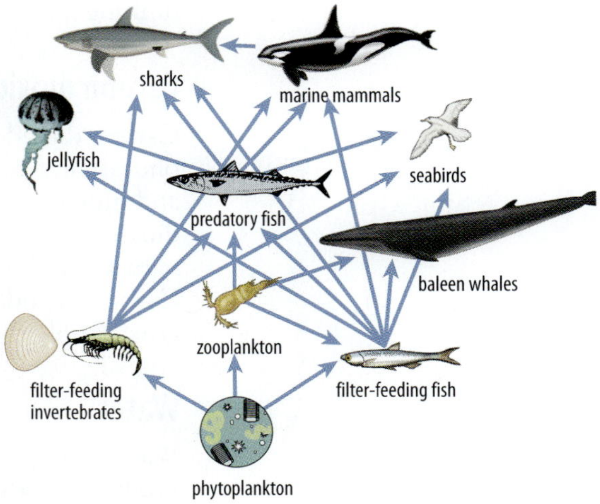

Figure 4. Marine food web.

1. Compare and contrast the two processes that autotrophs use to synthesize food.

2. What does the number of arrows leading *from* one organism to others suggest about that organism's importance in the food web?

3. What does the number of arrows leading *to* an organism suggest about that organism's likelihood of becoming endangered or extinct?

Resources for photosynthesis

Primary productivity levels vary with season and geographic location. You have identified and examined regions of extremely high and low productivity on land and in the ocean. These patterns of productivity are dictated by the availability of the resources necessary for photosynthesis. Autotrophs require carbon dioxide, water, sunlight, and nutrients to photosynthesize (Figure 1). When these resources are not present in adequate amounts for photosynthesis, they are referred to as limiting factors. Below, you will examine the likelihood of each resource being a limiting factor in productivity.

Carbon dioxide

Carbon dioxide is not likely to be a limiting factor in terrestrial or marine photosynthesis because it is plentiful in the atmosphere, as a result of respiration and human activities. As organisms convert food into energy, they release carbon dioxide (Figure 1). When humans burn fossil fuels like petroleum, coal, and natural gas, huge quantities of carbon dioxide are released into the atmosphere. Carbon dioxide is readily absorbed into the ocean from the atmosphere, so it is in good supply there as well.

Water

Water can often be a limiting factor for productivity in terrestrial environments. Productivity in deserts is typically very low because deserts lack the water needed to sustain all but the hardiest plants. In the ocean, obviously, water is never a limiting factor.

Sunlight

In the context of productivity, the sun's most important role is providing the light energy that drives photosynthesis. The availability of light to autotrophs varies, depending on latitude, season, and time of day. Therefore, there are places and times where sunlight is a limiting factor.

Nutrients

Organic or inorganic?

The terms **organic** and **inorganic** originally came from the idea that chemical compounds could be divided into two categories: those coming from or composed of plants or animals (organic), and those extracted from minerals and ores (inorganic).

Chemists now think of organic compounds as those that contain carbon. This definition works well for most compounds, but there are exceptions. For example, carbon dioxide is a carbon-based compound, but it is not considered organic.

Autotrophs use inorganic compounds containing nitrogen, phosphorus, potassium, calcium, silicon, and iron to build organic molecules. The concentrations of these **nutrients** vary throughout the environment. Nitrogen and phosphorus are particularly important because they are required in large quantities and play a critical role in growth and reproduction. Inorganic forms of nitrogen are the building blocks for amino acids, proteins, and genetic material (DNA, RNA). Similarly, phosphorus is an essential component of energy transport molecules (ATP), genetic material, and structural materials (bone, teeth, shell).

Despite the abundance of nitrogen gas in the atmosphere, it is often a limiting resource because it can be utilized by autotrophs only in certain forms. In the **nitrogen cycle**, nitrogen gas is converted to its most useful form, nitrate, through a series of reactions carried out by bacteria and fungi in the soil (Figure 5). Nitrates may be utilized by land plants, or may be carried to the ocean in groundwater or runoff and used by phytoplankton.

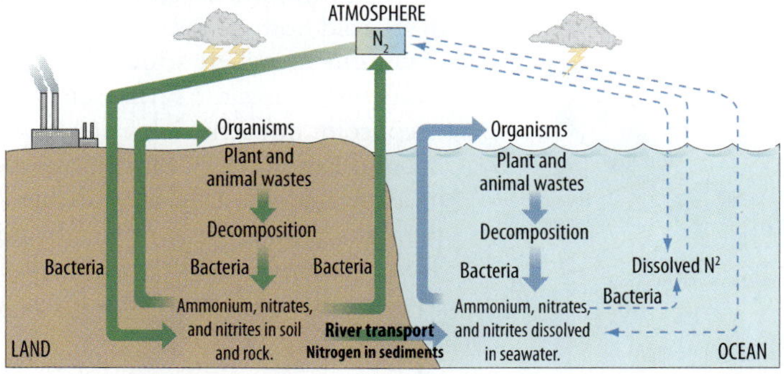

Figure 5. The nitrogen cycle.

Unlike nitrogen, phosphorus never exists as a gas. Phosphorus originates in rocks in Earth's crust in the form of phosphate salts, which are liberated from rocks by weathering (Figure 6). Phosphates are also an ingredient in some detergents and fertilizers. Precipitation carries phosphates into the soil, and runoff and ground water transport them to the ocean.

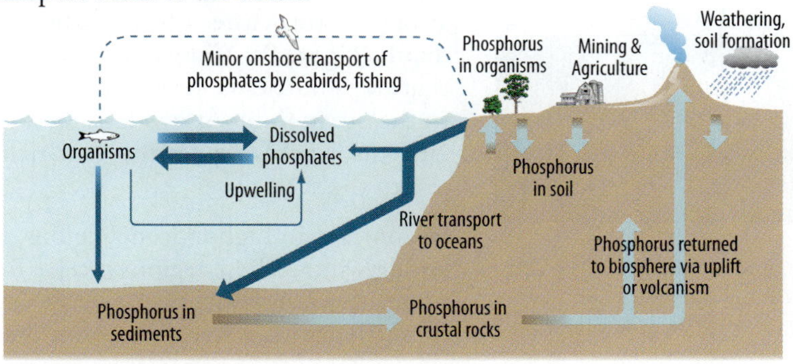

Figure 6. The phosphorus cycle.

4. Various human activities contribute additional nutrients to the ocean ecosystem. How might these additions influence marine productivity?

Sources of nutrients

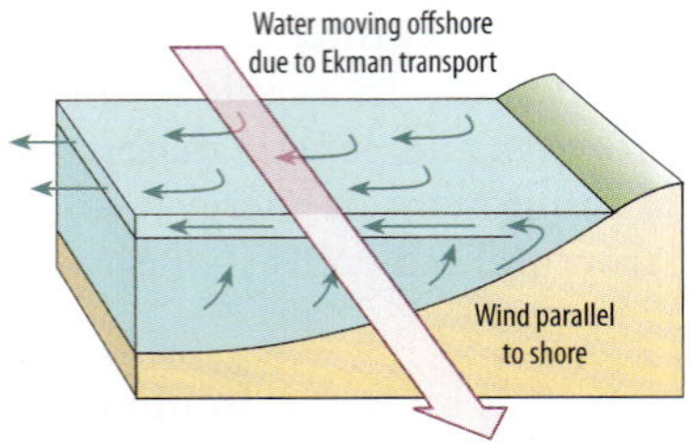

Figure 7. Coastal upwelling.

Most of the nutrients utilized by phytoplankton in surface waters originate on land. Nitrates and phosphates in the soil dissolve easily in water. These nutrients are easily leached or removed from the soil by precipitation and runoff, and are carried by groundwater, streams, and rivers to the ocean. Near the coasts, phytoplankton thrive on the nutrients entering the ocean from land, resulting in high productivity. However, large portions of the nutrients entering the ocean are not utilized and eventually sink to the ocean floor, accumulating in the sediment.

Not all nutrients enter the ocean from land. When marine plants and animals die they sink to the bottom, where decomposition liberates the nutrients, making them available for use again. However, except in shallow waters over the continental shelf, the nutrients on the ocean bottom can be utilized only when they are brought from the depths to the surface via **upwelling**, the upward movement of deep, cold bottom water to the surface. Coastal upwelling occurs in nearshore environments where strong winds blow parallel to the shore (Figure 7). **Ekman transport** causes the surface currents to deflect away from the shore, which pulls deep, cold, nutrient-rich water toward the surface. As you learned in Investigation 2.2 and Reading 2.3, Ekman transport is an offset between a current direction and its associated wind. In the Northern Hemisphere Ekman transport is deflected to the right, and in the Southern Hemisphere Ekman transport is deflected to the left. This phenomenon is caused by the Coriolis effect and by the slowing and deflection of water due to friction among successively deeper layers of water.

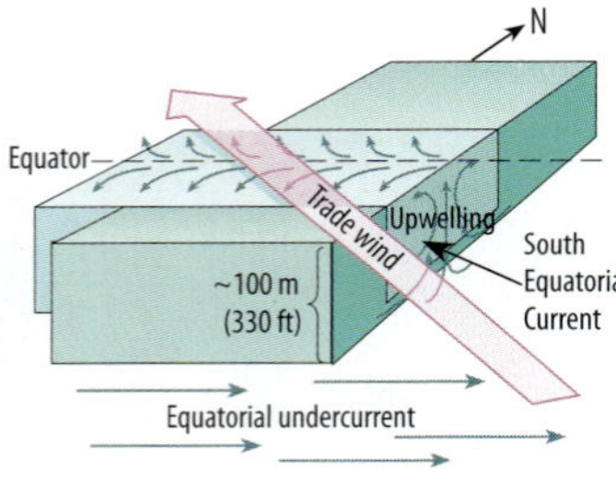

Figure 8. Equatorial upwelling.

Upwelling also occurs in the open ocean, near the equator. Equatorial upwelling is controlled by ocean currents—water moving westward on either side of the equator is deflected toward the poles and is replaced by cold, nutrient-rich water from below (Figure 8).

In the previous examples, upwelling is characterized by either winds or currents that move surface water and allow deeper water to "well up" in its place. The action of the wind or current is especially important in regions where there is a strong **thermocline**, a region where the water temperature changes rapidly with depth (Figure 9). Waters near the equator typically have a strong thermocline that traps nutrients in the cold, deep waters below, where they eventually sink to the ocean floor. At low latitudes, the mechanical action of strong winds or currents is necessary for nutrient-rich waters to penetrate the thermocline and rise to the surface.

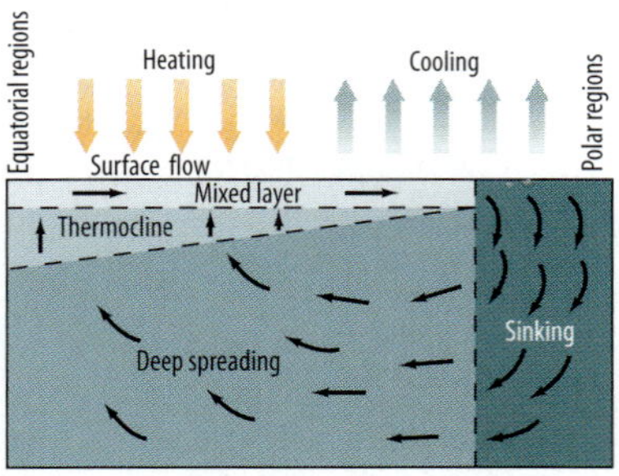

Figure 9. Schematic cross section of the ocean from equatorial to polar regions, showing change of depth to thermocline with changing latitude.

In contrast, in high-latitude regions, upwelling can occur without the help of strong winds. These regions receive little or no sunlight, resulting in a weak or nonexistent thermocline. As dense water sinks, it is replaced by an equal volume of nutrient-rich water rising from the adjacent depths.

5. What process makes nutrients in deep-ocean sediments accessible to marine phytoplankton drifting near the ocean surface?

6. Explain how the strength of the thermocline affects the availability of nutrients to primary producers.

7. Human activities can increase levels of nutrients in the ocean. Describe how human activities could influence the levels of *other resources* important in photosynthesis.

Too much of a good thing?

Previously, you observed that a shortage of one or more resources can decrease productivity. It may surprise you to learn that there can also be too much of a good thing when it comes to photosynthesis. For example, if the amount of solar energy received by the ocean were to intensify dramatically, tropical oceans could become too warm for photosynthesis to proceed properly.

Similarly, nutrients can be present in quantities so high that they become harmful. High levels of nutrients in coastal waters may cause a sharp increase in phytoplankton, which reduces light penetration, clogs the gills of marine organisms, and pollutes the water with waste products, some of which are toxic to other marine life. This may lead to **hypoxia**, a dramatic decrease in dissolved oxygen, or even **anoxia**, a total absence of dissolved oxygen, which can drive away mobile organisms and kill **sedentary organisms**.

sedentary organisms — organisms that are attached to a surface and cannot move freely.

Two conditions contribute to hypoxia in subsurface waters.

- **Stratification** or layering of the water column. Nutrient-laden runoff flowing into the ocean is less dense than salt water and floats on the surface. In summer, warm weather and calm seas inhibit mixing of the shallow, warm water with deeper, colder water. As a result, the oxygen from photosynthesis remains at the surface.

- **Increased decomposition** at depth, due to higher surface productivity. Phytoplankton die or are eaten by other organisms, creating large amounts of organic waste that sinks to the ocean floor. As the waste decomposes, nutrients are recycled but oxygen is also consumed, creating hypoxic conditions.

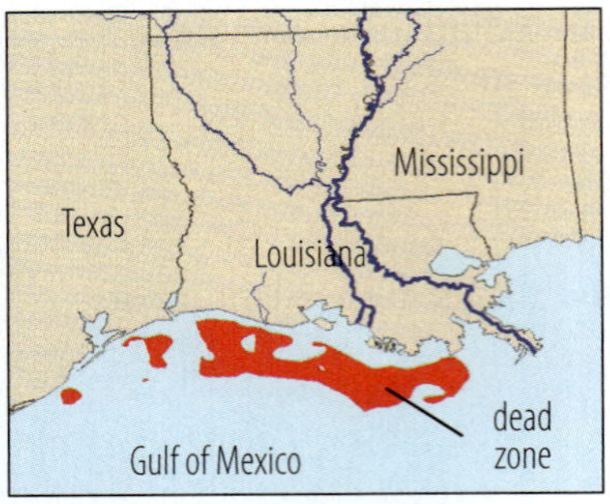

Figure 10. Location and extent of the 1993 Mississippi River dead zone.

The Gulf of Mexico is an important U.S. commercial fishery. In 2002, the Gulf accounted for 16.9 percent by weight (771,000 metric tons [850,000 U.S. tons]) and 24.7 percent by dollar value ($693 million) of the entire U.S. fish and shellfish catch. Each summer, a large hypoxic region known as the **Mississippi River dead zone** forms off the coast of Louisiana (Figure 10). The lack of dissolved oxygen in this region causes both the quantity and the diversity of economically important marine life to decrease dramatically, with potentially dire economic consequences.

8. Describe how an increase in nutrient levels can actually *lower* marine productivity in the marine ecosystem.

Is the Mississippi River dead zone unique?

In the United States alone, more than half of the estuaries experience hypoxia during the summer; up to a third experience anoxia.

9. Name two human activities that could result in an overabundance of nutrients being delivered to coastal waters. Explain your answer.

10. How might reduced light penetration affect primary producers as well as other animals higher up the food chain?

In the next investigation, you will examine changes in the Mississippi River dead zone from 1986 to 1993, and investigate patterns in the distribution of dead zones around the world.

Investigation 4.4

Dead zones

There are parts of the ocean where no fish swim, and where the bottom may be littered with the remains of bottom-dwelling crabs, clams, and worms. These areas, known as *dead zones*, pose a growing environmental concern. The causes of dead zones are complex and involve pollution in the form of excess nutrients from agriculture, industry, urbanization, and other sources.

Creating a dead zone

One of the largest dead zones in the world occurs in the northern Gulf of Mexico, where the Mississippi River flows into the open ocean off the coast of Louisiana. Starting in the 1950s, the waters of the Mississippi River began to show signs of elevated nutrient levels, particularly nitrogen compounds from agricultural runoff in the Mississippi River basin. At the same time, a dead zone began to form adjacent to the Mississippi delta in the Gulf of Mexico.

In this activity, you will look at the size of the Mississippi River dead zone over time, and compare it to dead zones on the east coast of the United States and around the world. This will help you to better understand how and where dead zones form, and to predict where they may develop in the future.

- Launch ArcMap, and locate and open the **etoe_unit_4.mxd** file.

Refer to the tear-out Quick Reference Sheet located in the Introduction to this module for GIS definitions and instructions on how to perform tasks.

- In the Table of Contents, right-click the **Mississippi River Dead Zone** data frame and choose Activate.
- Expand the **Mississippi River Dead Zone** data frame.

The Mississippi River dead zone

This data frame shows the Mississippi River and the Mississippi River *watershed*. The 31 states that are either partially or completely within the Mississippi River watershed boundary are shown in gray.

watershed — area of land that drains downslope toward the lowest point.

The Mississippi River watershed covers more than half of the 48 contiguous states, and many activities affect the quality of the water before it empties into the Gulf of Mexico. Next, you will examine how water is used, to identify possible sources of the excess nutrients in the Mississippi River dead zone.

- Turn on the **Water Consumption** layer.

This layer consists of pie charts for the 48 contiguous states, each showing the percentage of water consumed by six major water-use sectors (see sidebar). The overall size of each pie represents the total amount of water consumed by that state, and the slices represent the percentage of water used by each water-use sector.

Water-use sectors

- *Commercial* — facilities and institutions including hotels, restaurants, hospitals, and schools.
- *Domestic* — household use.
- *Industrial* — producing steel, chemicals, paper, plastics, minerals, petroleum, and other products.
- *Power* — steam-driven electric generators. (Does not include hydroelectric power.)
- *Mining* — extracting minerals, oil, and natural gas.
- *Agriculture* — raising animals and irrigating crops.

1. Which activities are the primary consumers of water in states within the Mississippi River watershed?

    a. Eastern portion of the watershed.

    b. Western portion of the watershed.

2. How might these activities contribute to nutrient enrichment of the Mississippi River?

Turn off and collapse the **Water Consumption** layer.

Turn on the **Precipitation** layer.

This layer shows average precipitation, in centimeters per year. Areas with low precipitation are shown in yellow and brown, and areas with high precipitation in blue.

3. Where is precipitation in the Mississippi River watershed

    a. highest?

    b. lowest?

Runoff from the land is one mechanism for transporting nutrients from farms and fields to the river.

4. Study the precipitation patterns, then predict where you would expect runoff to be highest in the Mississippi River watershed.

Turn off the **Precipitation** layer.

Turn on the **Runoff** layer.

This layer shows average runoff within the Mississippi River basin, in centimeters per year. Dark blue represents high runoff and light blue represents low runoff.

5. Which part of the Mississippi River watershed has the highest annual runoff?

6. Describe any relationships you observe between the precipitation and the runoff patterns.

7. How might levels of precipitation and runoff influence the size of the Mississippi River dead zone?

Turn off the **Runoff** and **Mississippi River System** layers.

Development of the Mississippi River dead zone

In this section, you will look at the extent of the dead zone at four different time periods, and compare the size of the dead zone to the climate data for the Mississippi River watershed during those four periods.

Click the QuickLoad button.

Select **Spatial Bookmarks**, choose **Gulf of Mexico**, and click **OK**.

In 1985, scientists began sampling water off the Louisiana coast to monitor the extent and characteristics of the Mississippi River dead zone.

Turn on the **Mississippi River Dead Zone** layer group.

Turn on the **1986 Dead Zone** layer.

Click the Identify tool.

In the Identify Results window, select the **1986 Dead Zone** layer from the list of layers.

Click inside the dead-zone to display the area of the dead zone for that year.

8. Record the area of the dead zone in Table 1. Round to the nearest 1 km^2.

Table 1 — Mississippi River dead zone data

Year	Area km^2	Area rank	Drought appearance
1986			
1988			
1990			
1993			

Turn off the **1986 Dead Zone** layer.

Repeat these steps using the **1988 Dead Zone**, **1990 Dead Zone,** and **1993 Dead Zone** layers. Remember to turn on each layer and select it in the Identify Results window. Record the area of each dead zone in Table 1. See Figure 1 for the location of the 1988 dead zone.

Close the Identify Results window.

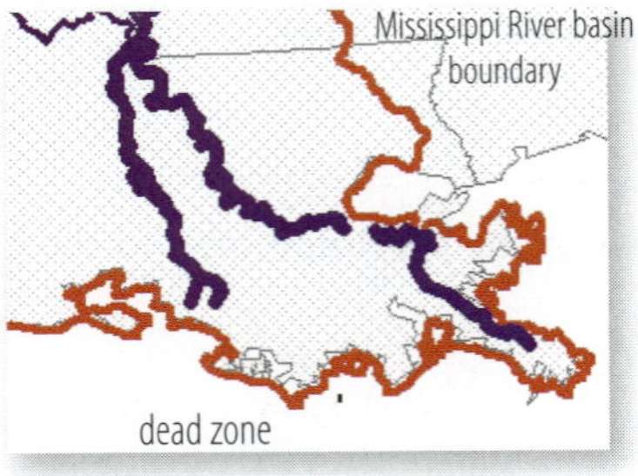

Figure 1. Location of 1988 Mississippi River dead zone.

9. In Table 1 on the previous page, rank the areas of the dead zones from 1 (largest) to 4 (smallest).

- Close the Identify Results window.
- Turn off all **Dead Zone** layers.
- Click the QuickLoad button.
- Select **Spatial Bookmarks**, choose **U.S. 48** , and click **OK**.
- Turn on the **Mississippi River Basin PDSI** layer group.
- Turn on the **1986 Drought Index** layer.

In the **Mississippi River Basin PDSI** layer group, green (index > 0) represents areas that are wetter than normal, and brown (index < 0) represents areas that are drier than normal.

10. In Table 1 on the previous page, describe the overall climate of the Mississippi River watershed for each of the four years, using one of the following descriptions: *very wet*, *wet*, *dry*, or *very dry*. Turn the four **Drought Index** layers on and off as needed to complete the table.

11. Using the data in Table 1, describe how the areas of the dead zones appear to respond to the climate conditions in the Mississippi River watershed.

12. Explain the changes in the area of the dead zone in terms of runoff and nutrients in the Gulf of Mexico.

The Mississippi River dead zone is not the only one in the U.S. or the world. Next, you will examine potential factors that may contribute to the formation of other dead zones.

Global dead zones

- Click the QuickLoad button.
- Select **Data Frames**, choose **Global Dead Zones**, and click **OK**.

The **Global Dead Zones** data frame shows the locations of dead zones throughout the world.

13. Describe the locations of the two primary clusters of dead zones.

Dead zones and population

Some scientists have noted that dead zones are related to the human populations of the adjacent continents. Many believe that dead zones may become the most significant human impact on oceans and ocean ecosystems in the 21st century. In this section, you will investigate this idea and the ways in which increasing population influences the size and number of dead zones.

14. If dead zones are related to areas with large populations, where on the map would you expect to find areas with large populations?

You can test your prediction by looking at the global population distribution, and see how it relates to the current distribution of dead zones.

💻 Turn on the **Global Population** layer.

How large an area is a 1° x 1° cell?

The spacing between latitude lines is fairly constant, but the spacing between longitude lines decreases away from the equator. Thus, the areas of 1° x 1° cells get smaller with increasing distance from the equator.

At the equator, a 1° x 1° cell has an area of about 12,500 km^2. At 60° latitude, the area of a 1° x 1° cell is only half that size, about 6200 km^2.

The **Global Population** layer shows the population for 1°× 1° *cells* (approximately rectangular regions within a grid) over Earth's surface.

15. Some scientists argue that high population causes dead zones. Examine the global patterns of population density and dead zones. Describe any observations that

    a. *support* (provide evidence in favor of) the idea that high population causes dead zones.

    b. *refute* (provide evidence against) the idea that high population causes dead zones.

There appears to be a connection between large populations and the formation of dead zones, but it is not a clear or consistent relationship. One possibility is that there is some important difference (other than size) between the populations adjacent to the regions where dead zones form and those near regions where dead zones do not form. Next, you will explore the possibility that *economic differences* among populations are a factor in the formation of dead zones.

💻 Turn off the **Global Population** layer.

💻 Turn on the **Gross Domestic Product** layer.

Gross Domestic Product (GDP) is the total value of goods and services produced by a nation within that nation's boundaries. Indirectly, it is a measure of the resources (energy, raw materials, people) available to and utilized by that nation.

Now you will determine the average GDP of countries adjacent to dead zones and compare it to the average GDP of countries without dead zones.

- Click the Select By Location button .
- In the Select By Location window, construct the query statement:

  I want to **select features from** the **Gross Domestic Product** layer that **are within a distance of** the features in the **Dead Zones** layer.
- Check the box to **Apply a buffer to the features in Dead Zones** and enter 100 kilometers.

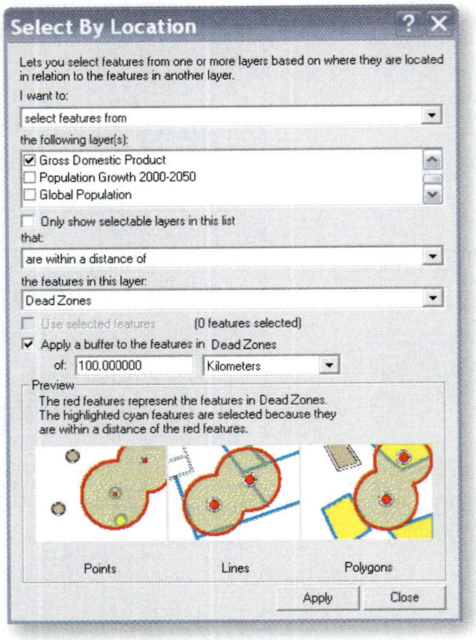

- Click **Apply**.
- Close the Select By Location window.

The countries that are near dead zones will be outlined. Next you will calculate statistics on the selected data.

- Click the Statistics button .
- In the Statistics window, calculate statistics for **only selected features** of the **Gross Domestic Product** layer, using the **GDP (billion $U.S.)** and **Population** fields. (Hold down the shift key to select multiple fields.)

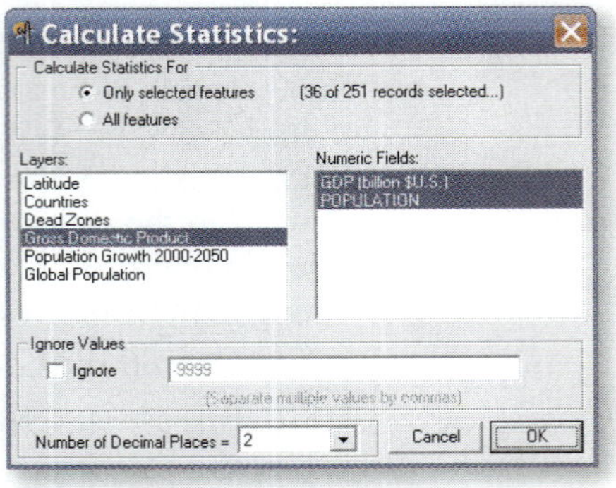

- Click **OK**.

In the statistics window, the number of countries adjacent to dead zones is given as the **Number of Records** and the GDP and population of those countries are given as the **Total**.

16. Record the number of countries (**Number of Records**), the GDP (**Total**), and the Population (**Total**) in Table 2. Record the values in the units shown in the table, to the nearest whole number for Total GDP and to the nearest 0.01 billion for Total Population.

Table 2 — Relationship of GDP per capita and presence of dead zones

	Number of countries	Total GDP *billion $U.S.*	Total population *billion*	GDP per capita
Near dead zones				
Not near dead zones				

- Close the Statistics window.
- Select the **Gross Domestic Product** layer.
- Click the Switch Selection button.

The countries that are *not* near dead zones will be outlined. Next you will calculate statistics on the selected data.

- Click the Statistics button.
- In the Statistics window, calculate statistics for **only selected features** of the **Gross Domestic Product** layer, using the **GDP (billion $U.S.)** and **Population** fields. (Hold down the shift key to select multiple fields.)
- Click **OK**.

In the statistics window, the number of countries not adjacent to dead zones is given as the **Number of Records** and the GDP and population of those countries are given as the **Total**.

17. Record in Table 2 the number of countries, GDP, and population values for countries not adjacent to dead zones.

- Close the Statistics window.

per capita [Latin] — for each person.

18. Calculate the GDP per capita both for the countries near dead zones and for the countries not near dead zones, and record them in Table 2. (Divide the Total GDP by the Total Population.)

19. How do the GDP per capita values differ for the two groups of countries?

20. How might GDP be a better indicator than population for the future development of dead zones?

Why use GDP per capita?

A large country may have a higher GDP than a smaller country simply because it has more people, even though the people in the smaller country may individually produce more goods and services. To correct for this, we use GDP per capita (i.e., per person) — the amount of goods and services produced by a country, divided by the number of people in that country.

In 2005, the world's population was about 6.1 billion people. By 2050, it is expected to increase to 8.9 billion, with the largest growth in countries having low GDP per capita.

21. If the total global GDP—the sum of all the goods and services produced by all nations—increases at a similar rate, what effect do you think it will have on the size and location of dead zones?

- Close the Statistics window.
- Quit ArcMap and do not save changes.

Wrap-up 4.5

Searching for solutions

The Gulf of Mexico is suffering from the effects of industrial and agricultural pollution, population growth, and urban development. The Gulf receives 1.1 trillion m^3 (300 trillion gallons) of runoff each year that contains a vast collection of pollutants originating from factories, hog-waste ponds, heavily fertilized farms, golf courses, and residential lawns, as well as oily grime from urban runoff.

Although these chemicals pose serious health hazards, the added nutrients pose the biggest problem, triggering a series of events leading to the formation of the Mississippi River dead zone. Excess nutrients have also been implicated in the deaths of coral reefs, decline of sea-grass beds, occurrence of red tides, and declining health of *estuaries* around the Gulf of Mexico.

estuary — body of water where a river meets the ocean, mixing fresh river water with ocean water.

What are red tides?

Red tides are caused by seasonal reproductive surges, or blooms, of certain species of marine algae. During a bloom, colored pigments in these tiny one-celled plants discolor the ocean surface, giving it a reddish-brown appearance.

Most of these algae species are harmless, but a few produce potent chemical neurotoxins (poisons that affect the nervous system). These toxins cause widespread fish kills, contaminate shellfish, and can be deadly to humans and other animals that eat contaminated seafood.

1. Describe the cascade of events that leads to the formation of a dead zone, beginning with the addition of excess nutrients. Be sure to discuss how the presence of excess nutrients affects primary producers and other consumers in the food web.

Although the Mississippi River dead zone is the largest in the United States, it is not the only one. Next you will determine the effect your community has on dead zones off the U.S. coasts.

- Launch ArcMap, and locate and open the **etoe_unit_4.mxd** file.

Refer to the tear-out Quick Reference Sheet located in the Introduction to this module for GIS definitions and instructions on how to perform tasks.

- In the Table of Contents, right-click the **U.S. Dead Zones** data frame and choose Activate.
- Expand the **U.S. Dead Zones** data frame.

This data frame shows the major rivers and watersheds in the contiguous United States. Find the approximate location of your city or town on the map.

- Using the Zoom In tool, locate and zoom in on the region you live in.
- Click the Identify tool.
- In the Identify Results window, select the **Major Rivers** layer from the drop-down menu.
- Next, click on the major river nearest to your town.

2. What major river carries runoff from your region to the ocean?

- In the Identify Results window, select the **Major Watersheds** layer from the drop-down menu.
- Click on the watershed in which your town is located.

3. What is the name of the major watershed in which your town is located?

- Close the Identify Results window.

Science can tell us a lot about how dead zones are created and their impact on the environment. However, cleaning up dead zones requires significant changes in our behavior as a society.

Search the Internet to learn more about the sources of pollution, the economic impact of dead zones, and the technological solutions to the problem; then answer the questions below. Some starting Web sites are provided on the next page.

4. Discuss how your city or town contributes to the formation of a dead zone. Think about local residential, commercial, agricultural, and industrial practices.

mitigate — to make less severe.

5. Discuss how you might mitigate the negative impact of some of these activities.

Media Viewer shortcuts

To open these Web pages from within ArcMap, click the Media Viewer button and choose the appropriate entry from the list of Web sites.

Web sites on the Gulf of Mexico dead zone

Exploring Solutions

Potential Solutions for Gulf of Mexico's "Dead Zone" Explored

Discusses the impact of the dead zone as well as ecological and technological approaches to reducing the nutrient flow into the ocean.

http://researchnews.osu.edu/archive/hypoxia.htm

Dead Zone Research

Environmental Literacy Council

Provides an overview of the problem and links to major government and research groups investigating the problem and solutions.

http://www.enviroliteracy.org/article.php/1128.html

Hypoxia

National Center for Appropriate Technology (NCAT)

Discusses causes of hypoxia and the human activities that contribute to it. Also discusses approaches to reducing nutrient flow to the ocean.

http://www.ncat.org/nutrients/map.html

Red Tides

The Gulf of Mexico Dead Zone and Red Tides

Discusses the effect of the dead zone on the quality and quantity of fish and seafood stocks in the Gulf of Mexico.

http://www.tulane.edu/~bfleury/envirobio/enviroweb/DeadZone.htm

Gulf in Peril

Deep Trouble — The Gulf in Peril

Series of articles in the Naples (Florida) *Daily News* outlining all the issues facing the Gulf of Mexico and potential solutions.

http://web.naplesnews.com/deeptrouble/index.html

Quit ArcMap and do not save changes.